THE SOUTH YORKSHIRE COALFIELD

A HISTORY AND DEVELOPMENT

Acknowledgements

I would like to acknowledge the support and assistance given by the staff of the Science, Technology and Management Department of the Birmingham Reference Library, which has been the principal reference source for the material used to compile this book.

I would also like to acknowledge the following: the staff of the National Coal Mining Museum for England, Caphouse Colliery, Yorkshire; John Goodchild, for access to material in The John Goodchild Collection, Wakefield, Yorkshire; The Coal Authority at Berry Hill, Mansfield, Nottinghamshire, for access to NCB records; Sheffield Archives, Sheffield, South Yorkshire; the Archives & Local Studies departments of Rotherham, Barnsley & Doncaster Libraries, South Yorkshire; Alan Rowles of Wales Bar, Sheffield; Keith Hopkins of Hoyland, Barnsley, South Yorkshire; John Powell of the Ironbridge Gorge Museum Library, Ironbridge, Shropshire; David Pollit, Mansfield, Nottinghamshire; Heiko Wenke, of Gevelsberg, Germany; Martin Walters, of Goldthorpe, South Yorkshire; and Frank Burgin, of Elsecar, South Yorkshire.

Front Cover Images
Top: Barrow Colliery
Bottom: Hickleton Main Colliery in the snow.

Back Cover Image
Cortonwood Colliery.

THE SOUTH YORKSHIRE COALFIELD

A HISTORY AND DEVELOPMENT

Alan Hill

To my Dad and brother, Barry – South Yorkshire coal miners.

TEMPUS

First published 2001

PUBLISHED IN THE UNITED KINGDOM BY:

Tempus Publishing Ltd
The Mill, Brimscombe Port
Stroud, Gloucestershire GL5 2QG
www.tempus-publishing.com

PUBLISHED IN THE UNITED STATES OF AMERICA BY:

Tempus Publishing Inc.
2 Cumberland Street
Charleston, SC 29401
(Tel: 1-888-313-2665)
www.arcadiapublishing.com

Tempus books are available in France and Germany
from the following addresses:

Tempus Publishing Group
21 Avenue de la République
37300 Joué-lès-Tours
FRANCE

Tempus Publishing Group
Gustav-Adolf-Straße 3
99084 Erfurt
GERMANY

British Library Cataloguing in Publication Data.
A catalogue record for this book is available from the British Library.

ISBN 0 7524 1747 9

Typesetting and origination by Tempus Publishing.
PRINTED AND BOUND IN GREAT BRITAIN.

Contents

Notes, Terminology and Abbreviations used in the text

Notes

1. In 1963 the NCB changed from a calendar year to a financial one, starting in April and finishing the following March.

2. In the financial year 1987-1988, beginning April 1987, the NCB changed to the metric system. From this date all output figures are quoted in metric tons. (Note that an imperial ton is equivalent to 1.016 metric tons).

3. On 1 January 1947 (Vesting Day) the British coal industry was nationalized, and from that date until 1994, when the industry was re-privatized, was owned by the nation.

4. Seam depths and thicknesses quoted can be very confusing, as these frequently vary significantly within the area worked by any individual colliery. Seam depths have generally been taken as those found in shafts during sinking. Sources such as *Sections of Strata of the Coal Measures of Yorkshire*, published by the Midland Institute of Mining, Civil and Mechanical Engineers in 1902-1913, and the Midland Institute of Mining Engineers in 1927, have been used. (Note that the lists of seams worked may not be fully comprehensive, and may not include all seams worked.)

Terminology

Beehive coke oven an early type of coke oven, shaped like a beehive. Beehive coke ovens were wasteful of heat and resulted in the entire loss of valuable by-products. Later coke ovens were more efficient, incorporating by-product recovery plant and waste-heat regenerators.
Bell pit an early form of coal (and metal) mining, employed where the seam was located close to the surface. A shallow shaft was sunk and the coal worked out from around its bottom until it was unsafe to work further. The cross-section of the shaft and working so formed resembling a bell.
Bind is a fine grained rock (mudstone) that accompanies coal seams.
Booster fan a fan, often placed underground, whose purpose is to supplement the main ventilating fan.
Brattice a partition, usually timber, placed in a shaft or drift to facilitate ventilation. The brattice divides the shaft into two routes, so that the flow of air can enter the workings down one side and return by the other. In the underground workings, brattice cloth (strong canvas, coated in tar or another substance to make it air-tight) was used to direct the airflow in the workings.
Bunker a large container, usually at the bottom of a shaft, to receive and store coal until it can be skip-wound to the surface.

Cage the frame, or lift, suspended in the shaft, which is used to raise and lower men, materials and coal. Cages often had more than one deck to increase their effectiveness.
Cardox shells a safer and widely used alternative to ordinary explosives, for use in coal mining. The system uses liquid carbon dioxide and is rechargeable.
Clod a soft, weak shale or clay, which frequently occurs as the roof of a coal-seam, or as partings in the seam.
Coal measures coal-bearing strata where coal-seams are to be found.
Coal plough a type of coal-cutting machine used in longwall mining. It 'planes' a narrow strip of coal from the face, allowing it to fall onto a conveyor. Developed in Germany in the 1940s, these were popular in the thin, hard seams of South Wales and Yorkshire.
Coal preparation plant surface plant used to wash and grade coal before despatch.
Concealed coalfield the part of a coalfield where the coal measures are overlain by newer strata, typically the magnesium limestone from the Permian age.
Dip the direction in which the strata and coal-seams increase in depth from the surface.
Downcast shaft following the Hartley Colliery disaster of 1862, legislation stipulates that collieries must have two means of entering or leaving their workings. In the majority of cases this means two shafts. Having two shafts also facilitates ventilation. One shaft is the downcast, where air enters the workings, and the other is the upcast shaft, where stale air leaves the workings.
Downthrow a fault in the earth's crust which has resulted in the strata, and the accompanying coal-seams, being at a lower level than the adjacent strata.
Drift a tunnel in, or leading to, a coal-seam. Cross-measure drifts are inclined tunnels that connect coal-seams. Surface drifts are tunnels from the surface to the coal-seam(s).
Exposed coalfield that part of the coalfield where the coal measures are not overlain by newer strata, so the coal measures outcrop at the surface.
Fault a geological dislocation in the strata, resulting in the strata being discontinuous.
Gannister a hard siliceous fireclay, found with some coal-seams, which was used to manufacture high-quality firebricks.
Gate (or **gate road**) see Maingate.
Goaf (gob in South Yorkshire) the space left after the removal of the coal. It is supported by packing with stone and waste material, or is allowed to collapse in a controlled manner.
Hand-got/won coal got (won) by manual methods, such as pick and shovel.
In-bye towards the coalface, and away from the shaft.
Intake road the tunnel that carries fresh air from the downcast shaft into the colliery.
Koepe (friction) winder a winding system which uses the friction between the winding rope(s) and the drive pulley(s). Popular in Germany, where it originated, and later in the UK.
Level a horizontal or slightly sloping drift (see **drift**), sometimes used for drainage purposes, or driven in the side of a hillside to meet a coal-seam.
Longwall mining a method of coal mining in which coal is worked along a long face, or wall, running between two parallel roadways (tunnels).
Main in South Yorkshire the suffix 'main' to a colliery name, e.g. Manvers Main, signified that the colliery worked the 'main' or Barnsley Bed Seam.
Maingate a gate, or gate road, is a tunnel which serves each end of a longwall coalface. The maingate is where the fresh air enters the coalface, the tailgate is the exit tunnel for the air.

Merry-go-round system introduced in the late 1960s. This system enabled a train of coal wagons to be loaded rapidly without stopping and return, with minimum delay, to a power station or similar high tonnage user of coal.
Mine-car a larger capacity form of tram (a tram typically held up to one ton, a mine-car five tons, of coal) suitable for locomotive haulage underground.
Opencast a method of mining coal similar to quarrying.
Outcrop where a coal-seam reaches and breaks out at the surface.
Out-bye towards the shaft, and away from the coalface of the colliery.
Paddy mail the name given to a manriding wagon, used underground to transport men to and from their place of work.
Pillar and Stall a system of working the coal-seam where pillars of coal are left to support the roof. It has been superceded by the longwall system, in which the whole of the seam is mined.
Pit a common term for a shaft, or a colliery.
Power loading a method of mining where the cutting machine cuts the coal from the coalface and loads it onto a conveyor running along the face.
Regenerative coke oven coke ovens in which the waste heat from the oven is reused by the use of regenerators.
Return road the tunnel along which the stale air travels back to the upcast shaft.
Retreat mining mining in which the roadways are driven to the extremity of the royalty and the coal mined back towards the shafts. In 'conventional' mining, coal is mined from the shafts outwards. Retreat mining requires a larger initial investment, but enables very high output rates to be achieved and sustained. Employed successfully in the high output drift mines sunk from the late 1960s in South Yorkshire.
Rise the direction in which the strata, and coal-seams, reduce in depth from the surface.
Roadway (or **road**) another name for an underground tunnel.
Royalty the minerals and the right to work them, which belong to the owner of the freehold. The ownership for the surface could be vested in one person and for the mines in another. Until the Coal Act of 1938, ownership for coal deposits remained with the proprietor of the land, who charged royalties to the mine operators. The 1938 Coal Act transferred ownership for coal deposits to the state and provided for payment of compensation to the landowners.
Screen a device with fixed, or vibrating, bars or perforated steel sheets, which grades coal by size.
Shaft guides fixtures in a shaft used to guide the cages running in the shaft. In early British coal mines these were rigid timber or steel rails. Later, iron and steel ropes were used.
Shaft pillar the block of coal left beneath a colliery to support the shaft and colliery surface buildings. The pillar of coal is always significantly larger in area than the surface area of the colliery. The only material removed from the shaft pillar would be to form the roadways into the workings and, in these, the roof would be strongly supported by a brick or concrete lining.
Short wall working a technique for working coal employing a longwall face which is kept deliberately short (typically 150-200yds in length).
Skip a container (a form of large bucket) in which coal is carried up the shaft.
Slack is generally understood to mean small coal.
Spontaneous combustion some coal-seams are liable to spontaneously ignite and cause underground fires, with the allied risk of explosion. Adequate ventilation and attention to safety procedures minimise this risk. One seam prone to spontaneous combustion was the Barnsley Bed.

Staple shaft a shaft sunk entirely underground, between two or more seams, to facilitate the movement of materials, men or coal between the seams, or provide a ventilation circuit in the underground workings.
Steam-coal a grade of coal that is suitable for steam-raising in a boiler. The principal characteristics being that the coal; burns easily, produces a lot of heat and a minimum of ash and clinker.
Stowing the process by which stone and spoil are used to pack the goaf, and other voids. Spoil is usually transported to the working faces, from the surface washery, pneumatically.
Tailgate see Maingate.
Take a term commonly used to mean the same as Royalty, the area of coal worked.
Tippler a device for emptying trams, usually a rotary tippler, emptying the trams into a bunker or railway wagon located beneath it.
Trepanner shearer a coal-cutting machine used in longwall mining, with a rotating head and cutter placed around its circumference. The trepanner feeds onto a conveyor. Introduced into British mining in the 1950s.
Throw is the displacement between layers of strata caused by a fault.
Trough a geological term meaning an area of coal-seams which are lower than the surrounding strata because they have been 'downthrown' by the faults surrounding them. Because the coal-seams were deeper than in the surrounding area, these areas were usually worked at a later date, such as at Frickley and Maltby.
Tubbing a casing inserted into a shaft, as it was being sunk, to keep back the water. Expensive and only used where the ground was wet or 'soft'. In the nineteenth century, cast iron tubbing sections were most generally used.
Upcast shaft see Downcast shaft.
Washout an area in which the coal-seam has been eroded away.

Abbreviations

CEGB – Central Electricity Generating Board.
LMS – London, Midland & Scottish Railway
LNER – London, North Eastern Railway
m – million, e.g. 1.2m tons (1,200,000 tons).
NCB – National Coal Board
NUM – National Union of Mineworkers.
OMS – Output per man-shift, a commonly used performance measure of the nationalized coal industry.
RCHME – Royal Commission on the Historic Monuments of England.
ROLF – Remotely operated longwall face.

The South Yorkshire Coalfield is situated in the Midland Coalfield approximately bounded by the triangle formed by Barnsley, Doncaster and Sheffield.

Introduction

In the book *Schumacher on Energy* (a collection of speeches and writings by Dr E.F. Schumacher, edited by Geoffrey Kirk) the author says: 'There are only two basic items in the world economy – food and fuel. All the rest are secondary… there is no substitute for energy; the whole edifice of modern life is built upon it.'

This book has been written to provide an overview of the history and development of the South Yorkshire Coalfield, showing the important contribution that it has made to the energy needs of Britain. Geologically the South and West Yorkshire coalfields are two separate (but joined) coalfields, formed by separate coal swamps during the Carboniferous period.

For the purposes of this book, the South Yorkshire Coalfield has been defined as the region of the Yorkshire Coalfield south of a line from a little north of Barnsley, stretching from Woolley to Askern, as shown on the map on page twelve.

After nationalisation of the coal industry, in 1947, the Yorkshire and South Yorkshire regions of the coalfield were reorganized on a number of occasions, the boundaries changing to eventually incorporate collieries such as Shireoaks, Steetley, Manton and Firbeck. These collieries are geographically sited in Nottinghamshire, as such, they have not been included in this work.

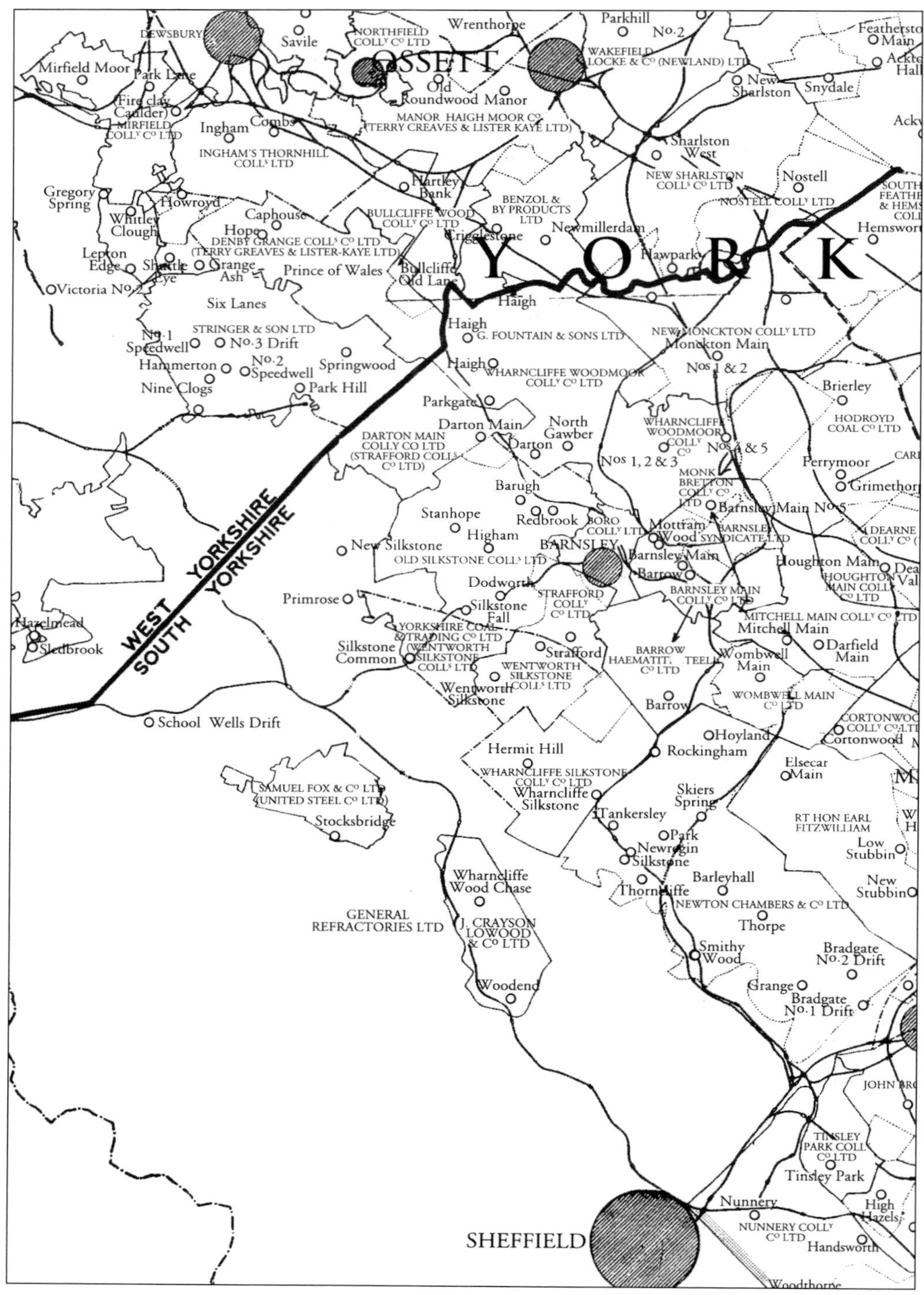

The division of the Yorkshire Coalfield into the distinct South & West Yorkshire Coalfields, and areas worked by individual colliery companies.

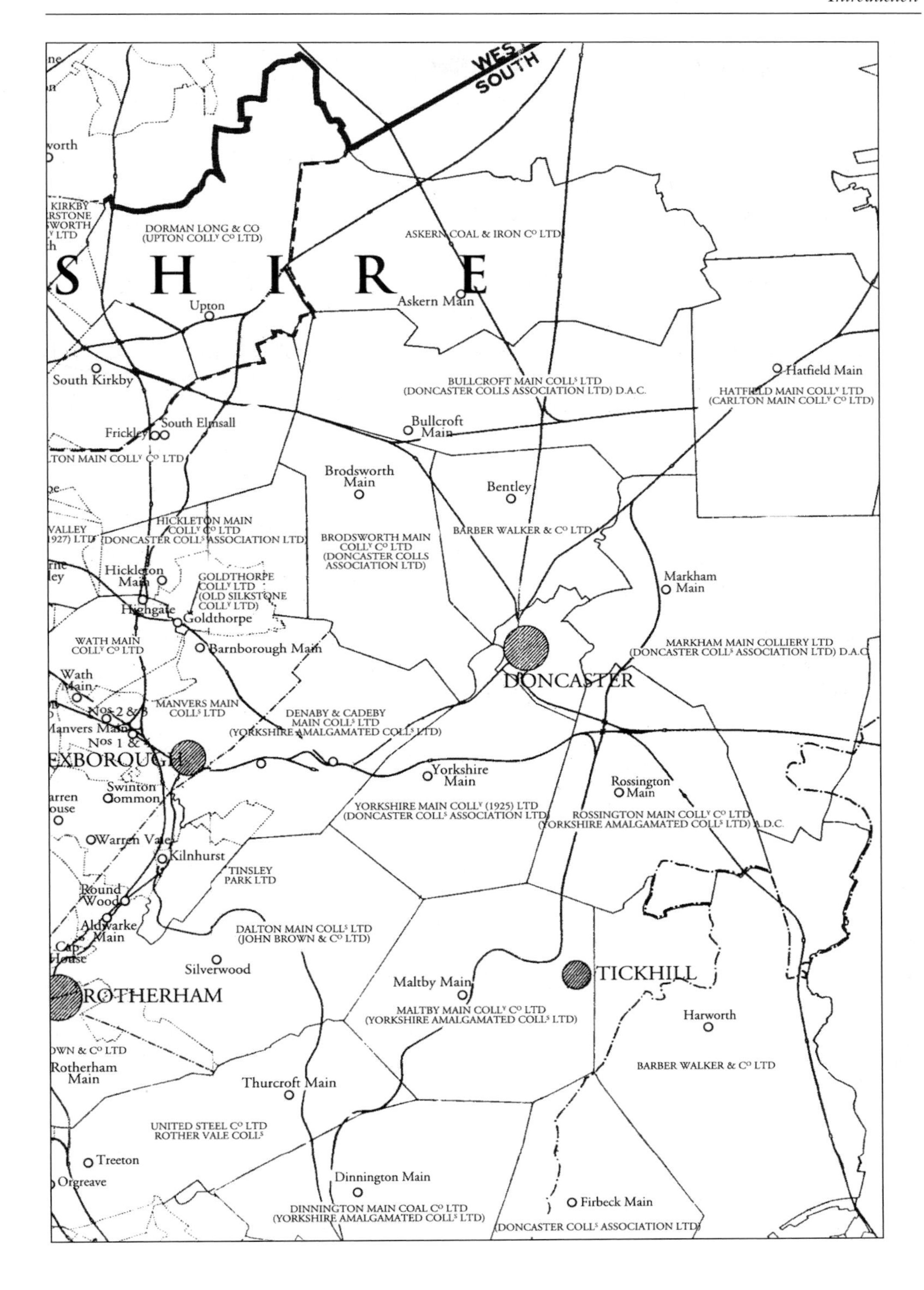
WEST
SOUTH
SHIRE
DORMAN LONG & CO (UPTON COLLY Cº LTD)
ASKERN COAL & IRON Cº LTD
Upton
Askern Main
South Kirkby
BULLCROFT MAIN COLLS LTD (DONCASTER COLLS ASSOCIATION LTD) D.A.C.
Hatfield Main
HATFIELD MAIN COLLY LTD (CARLTON MAIN COLLY Cº LTD)
Frickley
South Elmsall
Bullcroft Main
Brodsworth Main
Bentley
HICKLETON MAIN COLLY Cº LTD (DONCASTER COLLS ASSOCIATION LTD)
BRODSWORTH MAIN COLLY Cº LTD (DONCASTER COLLS ASSOCIATION LTD)
BARBER WALKER & Cº LTD
Hickleton Main
GOLDTHORPE COLLY LTD (OLD SILKSTONE COLLY LTD)
Markham Main
Highgate
Goldthorpe
WATH MAIN COLLY Cº LTD
Barnborough Main
MARKHAM MAIN COLLIERY LTD (DONCASTER COLLS ASSOCIATION LTD) D.A.C.
Wath Main
DONCASTER
MANVERS MAIN COLLS LTD
DENABY & CADEBY MAIN COLLS LTD (YORKSHIRE AMALGAMATED COLLS LTD)
Yorkshire Main
Swinton Common
Rossington Main
YORKSHIRE MAIN COLLY (1925) LTD (DONCASTER COLLS ASSOCIATION LTD)
ROSSINGTON MAIN COLLY Cº LTD (YORKSHIRE AMALGAMATED COLLS LTD) A.D.C.
Warren Vale
Kilnhurst
TINSLEY PARK LTD
Round Wood
Aldwarke Main
DALTON MAIN COLLS LTD (JOHN BROWN & Cº LTD)
Silverwood
Maltby Main
TICKHILL
ROTHERHAM
MALTBY MAIN COLLY Cº LTD (YORKSHIRE AMALGAMATED COLLS LTD)
Harworth
BARBER WALKER & Cº LTD
Rotherham Main
Thurcroft Main
UNITED STEEL Cº LTD ROTHER VALE COLLS
Treeton
Orgreave
Dinnington Main
DINNINGTON MAIN COAL Cº LTD (YORKSHIRE AMALGAMATED COLLS LTD)
Firbeck Main
(DONCASTER COLLS ASSOCIATION LTD)

1
Overview of The South Yorkshire Coalfield

The South and West Yorkshire Coalfields

The Yorkshire Coalfield is divided into two distinct regions along a line a little north of Barnsley, stretching from Woolley, to Askern, north of Doncaster. Along this important zone of change the Barnsley Seam, the principal coal-seam of Yorkshire, changes from a high-quality seam over 6ft in thick ness, to one of inferior quality. Geologically, this reflects the fact that the two regions were formed in separate and distinct coal-swamps, separated by a barrier of land at the time of their formation. The difference between the two coal basins is further emphasised by the difference between the coal-seams. Many of the important coal-seams of South Yorkshire become insignificant in the West Yorkshire Coalfield. The focus of this history is the South Yorkshire Coalfield.

History & Development

The coal measures of South Yorkshire dip gradually from their outcrop in the foothills of the Pennines in the west, as they move eastwards in the direction of Doncaster. The coal-seams of Yorkshire dip gradually to the east and south-east. In the east the seams are overlain by later Permian and Triassic rocks. This part of the coalfield is referred to as the concealed coalfield. The coalfield was formed during the Upper Carboniferous Period and the economic coal-seams are spread through what are known as the Lower and Middle Coal Measures. The Middle Coal Measures, between the Silkstone and Barnsley Seams, contain the most important coal-seams of the coalfield. It was to the Barnsley Bed Seam that South Yorkshire owed its reputation as a coal-producing region, as this seam exceeded all others in both thickness and quality.

The exposed coalfield in the west is terminated by a north-south scarp, which coincides with the exposure of the Magnesian Limestone, just to the west of Doncaster. To the north-east of Doncaster, particularly in the Bentley district, the low-lying area was formerly the bed of a large lake. Over wide areas the land is less than 25ft above sea level, and the risk of flooding in this area has been increased by mine subsidence.

The wealth of the South Yorkshire Coalfield was derived from its extent rather than the vertical density of its coal-seams. The seam density is greatest in the central region of the coalfield decreasing westwards, due to the outcropping of the seams, and eastwards, due to thinning and deterioration of the seams. For these reasons the reserves in the central region, where coal mining had been established for half a century or more, were often as great as in the unworked areas in the concealed part of the coalfield to the east.

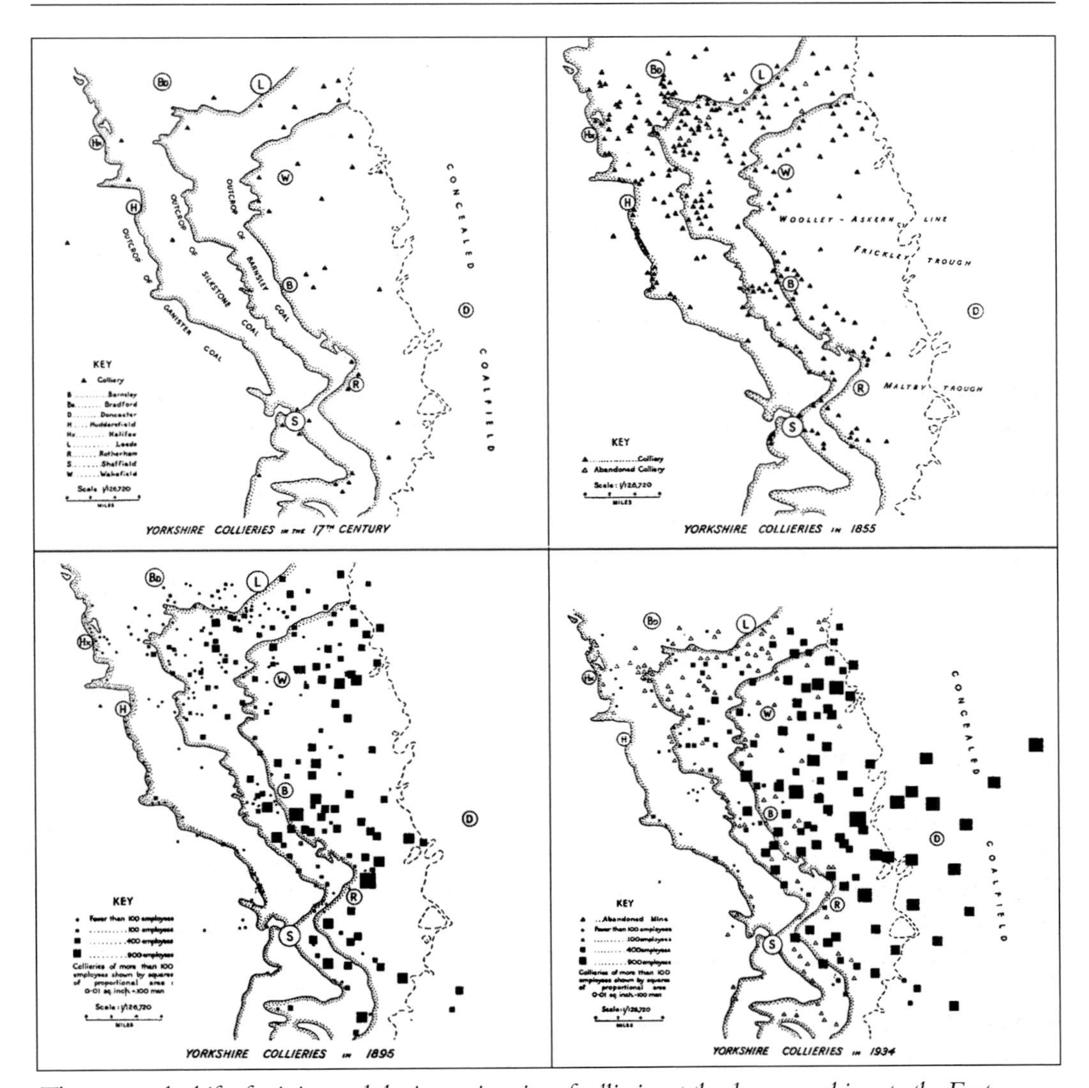

The eastwards shift of mining and the increasing size of collieries at the deeper workings to the East.

The structure of the coalfield was quite simple, with the seams dipping gently to the east. There was however one important structural disturbance – along the middle section of the river Don, between Sheffield and Mexborough, where a series of north-east to south-west faults, called the Don Faults, occurred. On either side of this faulted zone were two troughs, the Frickley Trough to the north and the Maltby Trough to the south. The troughs, where the coal-seams were deeper, were to account for the late development of these areas of the coalfield. For collieries sunk within the main zone of the faults, such as Aldwarke Main and Kilnhurst, this faulting had an important bearing on coal working.

The chief coal-seams of South Yorkshire were the Barnsley, Parkgate, Thorncliffe, Haigh Moor and Silkstone Seams. The historic development of the coalfield commenced from the outcrop of the seams in the foothills of the Pennines to the east. There is evidence that coal was mined sporadically during the Roman occupation. This was probably along a narrow arc along the outcrop from Rotherham to the west of Barnsley and onwards to Wakefield.

Coal was being worked at Silkstone, to the west of Barnsley, and at Masborough, close to Rotherham, in the thirteenth century. During the sixteenth and seventeenth centuries there are frequent references to the working of coal in the neighbourhood of the towns of Barnsley, Rotherham and Sheffield.

Improvements in transport and engineering resulted in the exploitation of most of the exposed coalfield by the middle of the nineteenth century. Further improvements in mining technology, combined with the gradual exhaustion of the shallower seams in the western part of the coalfield, enabled and provided an incentive for mining further to the east and away from the outcrop.

The location of collieries after about 1860 was largely dictated by the presence of railways, with many collieries sited on or close to main lines. The increasing tonnage of coal being raised required a more efficient form of distribution.

Prior to 1850 the collieries had shallow workings. The early workings were clustered along and close to the outcrop along the western margin of the coalfield. Coal would have been mined from shallow drifts, bell pits and shallow shafts. Few shafts would have exceeded 300ft in depth. As these collieries were working coal at shallow depth, it was more economic to sink pits and drifts close together, working a limited area of coal around each before moving on to sink again. The technology, in terms of ventilation of the workings of the pit and haulage of the coal underground, would have been relatively primitive, limiting the area of coal worked.

As the development of the coalfield extended eastwards, and the working depth of the collieries increased, then the cost of sinking and mechanical equipment increased dramatically and technical difficulties associated with ventilation and drainage became more apparent. In order to recoup the increased investment, it became necessary to work more extensive royalties and produce greater outputs. In the west of the coalfield collieries would typically work areas of 1,500 acres and seams at 100 to 200yds depth. In the central region of the coalfield, 3,500 acres and working depths of 250 to 600yds were typical. In the east, royalties up to 10,000 acres and shafts of 800 and 900yds depth were not uncommon.

The relentless movement towards the east, and the large-scale mining of the 'fiery' Barnsley Bed Seam from the middle of the nineteenth century, led to an increasing number of mining tragedies in the South Yorkshire Coalfield, culminating in the Barnsley Oaks Colliery disaster of December 1866, when 361 perished. The disaster remained the worst in British mining history until the Senghenydd Colliery disaster of 1913.

Before the 1850s a large number of small collieries and drift mines were sunk near to the outcrop down the western margin of the coalfield. By the 1870s and 1880s the centre of the coalfield was being rapidly developed. The final surge in mining development came in the first twenty years of the twentieth century with the development of the Doncaster area of the coalfield. This resulted in the rapid rise in output between 1905 and 1913, and Doncaster changing from a small agricultural town to a large industrial one.

Growth and Expansion of Output

Outputs and persons employed (taken from the Inspector of Mines Reports):

	South Yorkshire		**Totals for Yorkshire**	
Year	**Output**	**Number employed**	**Output**	**Number employed**
1851	*3,375,000*	*11,750*	*6,750,000*	*23,500*
1860	*4,640,000*	no record	*9,284,000*	no record
1870	*5,770,000*	*18,500*	*11,545,400*	*36,500*
1880	*9,260,000*	*31,000*	*17,473,806*	*60,474*
1890	*12,290,000*	*41,000*	*22,338,886*	*76,797*
1900	*16,100,000*	*55,000*	*28,247,249*	*100,826*
1905	*17,360,000*	63,005	29,923,654	112,640
1910	23,822,973	87,400	38,300,600	146,956
1913	28,255,294	99,834	46,671,243	161,220
1915	25,900,619	86,385	40,343,694	140,753
1920	22,170,030	105,429	36,166,444	173,474
1924	30,931,671	122,582	46,568,688	195,326

(note: figures in italics are estimates)

The boom effect of the development of the huge Doncaster area collieries is dramatically shown by the jump in output from the 1900s, together with the increasing importance of the South Yorkshire area from this period. The output of the district increased from an estimated 3.3m tons per annum, or 6 per cent of UK output, to almost 31m tons, or 11½ per cent of UK output, in 1924, representing over 85 per cent of the output of Yorkshire.

The Royal Commission on Coal Supplies of 1904 showed estimates of the reserves of coal in seams of 12in, and more, comprised 8,367,385,600 tons in West Yorkshire and 10,770,620,700 tons in South Yorkshire.

Technology

As the depth of mining and outputs increased, the need for more powerful winding-engines became apparent. During the period of maximum expansion of the coalfield, between 1890 and 1930, Messrs Markham & Co. of Chesterfield played an important role in providing the steam winding-engines to fulfil the needs of the industry. Of the 112 steam winding-engines built by Markham & Co. between 1895 and 1930 inclusive, forty-one were installed at South Yorkshire collieries. Of these engines; ten were in the newly opened Doncaster area of the coalfield, and the remainder were principally for re-equiping collieries in the central region, such as Manvers Main and Hickleton Main, which were increasing their output or sinking to deeper seams. Three of these engines, those installed at Grimethorpe, Silverwood and Frickley Collieries; were among the largest steam winders ever built by Markham.

Transport

The River Don was first made navigable and used for the carrying of coal in 1734. Further canals were constructed, completing the network of canals in the South Yorkshire Coalfield by 1819, with access to the centre of Sheffield. By 1835 159,000 tons of coal were brought to Sheffield by canal from collieries in the neighbourhood of Tinsley and Rotherham, to the north of Sheffield.

The first railway between Rotherham and Sheffield opened in 1838 and the first main line railway from Derby to Leeds, via Chesterfield and Rotherham, was constructed between 1838-1840. Further railway lines were built and eight important routes had been completed in South Yorkshire by 1855 and by 1902 a total of fourteen railway routes criss-crossed the coalfield.

The coming of the railways provided a further boost to the development of the coal industry, the network spreading rapidly throughout the coalfield. All major collieries, particularly those sunk from the 1870s onwards, were sited on main line routes, or were connected to them by short branch-lines.

Coal Carbonization In South Yorkshire

The carbonization of coal to produce coke has been an important industry in South Yorkshire since the beginning of the twentieth century. The abundance of seams yielding high-quality coking-coal providing the principal impetus to the industry. Prior to 1900 coal was carbonized in beehive type coke-ovens, generally without by-product recovery. A notable exception was the pioneering work carried out by an analytical chemist, called J.H. Worrall, at the Thorncliffe works of Messrs Newton, Chambers & Co. Newton, Chambers had already recovered ammonium sulphate and light oils from their patent beehive coke oven, and soon became well known for the manufacture of a disinfectant sold under the trade name *Izal*.

The manufacture of coke in a retort, or closed oven, with by-product recovery, began in South Yorkshire in 1900, when the Simon Carves Co. constructed a battery of ovens at Wharncliffe Silkstone Colliery. Following the successful introduction at Wharncliffe Silkstone there was a growth in the number of such plants in South Yorkshire and Derbyshire.

By the end of the first decade of the twentieth century, the production of coke in Britain had reached an output of 10m tons. Yorkshire was the second largest producer, after Durham, with over twenty-five per cent of the output. The expansion in the use of by-product ovens, and rapid development of the technology, led to the proliferation of manufacturers and contractors.

Despite the introduction of more advanced coke manufacture and by-product recovery, many beehive coke ovens were in use during the early twentieth century. The last beehive coke ovens in South Yorkshire were at Hazelhead Colliery, surviving until nationalisation of the coal industry in 1947.

John Brown & Co. built a notable coking plant at Cnaklow, Rotherham. The first two banks of waste-heat ovens were built in 1902 and 1904. In 1927 the company announced that they had ordered twenty-eight underjet coke ovens, capable of carbonizing 3,000 tons of coal per week. The plant closed in November 1962, long after the parent colliery had closed in 1954. The plant provided a valuable source of income for thirty-three years and was by no means worn out when closure finally came.

The coking plant at Manvers Main Collieries was special in a number of ways. The first plant was erected in 1906, and comprised some of the earliest Koppers' regenerative ovens. The plant was progressively extended, until 1932, when the directors of the company decided to dismantle the whole of the existing plant and install an up-to-date facility comprising thirty underjet ovens, followed by a further fifteen in 1934. The plant was extended in 1957 with the addition of a further sixty-six ovens. The existing ovens were rebuilt and the plant generally refurbished and extended. The result was the largest coking plant in Europe, built to process the coal, from four collieries, at a central coal preparation plant located at Manvers.

An interesting coking plant was built in 1929 for the Nunnery Colliery Co. at Handsworth, near Sheffield. It was unusual as it was the only plant in the district built specifically to produce domestic smokeless fuel, rather than blast-furnace coke. In the 1950s, when the NCB were making strenuous efforts to improve the quality of smokeless fuel, it became the first plant to be chosen for a series of full-scale experiments. This plant closed in 1961.

It is a feature of South Yorkshire collieries that many of them were part of composite undertakings. They were associated with coke-ovens, by-product plants, brickworks and in some cases blast furnaces and steel works. A large quantity of the surplus coke-oven gas was fed into the South Yorkshire coke-oven gas grid, which was created in the 1930s.

Markets For The Coal

Until the mid-nineteenth century, the market for South Yorkshire coal was almost entirely within the local region. It kept pace with the industrial development that was taking place in Yorkshire and neighbouring counties. The growing iron and steel industries of Sheffield and Rotherham, and the textile industry of the West Riding, absorbed increasing amounts of industrial fuel. The expanding population took household fuel and gas-coal, and the railways took large quantities of steam-coal.

South Yorkshire coal was high grade, its steam-coal was second-to-none, and had a world-wide reputation. As well as being exported, it was heavily used by steamship lines, railway companies, manufacturing industry and for power generation over the whole of the eastern and southern counties of England.

South Yorkshire gas-coals were used throughout the country and were mixed with poorer quality gas-making coals from other mining districts. Household coal was produced and South Yorkshire coking coal was probably the finest in the country.

In 1875 the Yorkshire Coalfield as a whole exported about six per cent of its total output. With the eastward development of the South Yorkshire Coalfield, particularly the large collieries of the Doncaster area, that were closer to the ports on the east coast, the export trade gradually increased. In 1913 the export trade of the Yorkshire coalfield had increased to twenty per cent of the total output. The large collieries sunk in the east were sunk in the anticipation of shipping fifty per cent of their output. Before the First World War there was an important export market to the Baltic.

2
Coal-Seams Worked and Geology

Overview of the Coalfield

The Yorkshire Coalfield covers a large area of Yorkshire bounded by Leeds, Bradford, and Halifax in the north-west, Sheffield in the south-west and extends eastwards beyond Doncaster. It forms the northern-end of the Great Midland Coalfield, which covers an area of some 900 square miles (the Yorkshire portion covers about 500 square miles). This geologically continuous coal basin stretches from Yorkshire down through Derbyshire and Nottinghamshire. The same classes of coals, with different local names, occur throughout the coalfield. The three regions formed a single massive coal basin, approximately twice the size of the South Wales Coalfield, sharing a common geological structure and sequence of coal-seams.

The western boundary of the Yorkshire Coalfield is marked by the outcrop of the coal measures on the eastern flank of the Pennine uplift. A belt of east-west faults to the north of Leeds limits the exposed coalfield in the north. The principal faulting structures within the coalfield are associated with the Pontefract and Don Valley systems of faults. As one crosses the coalfield from west to east, the outcrops of the various seams are encountered in ascending order until we pass from the exposed coal measures to the concealed coalfield, where the Carboniferous rocks lie beneath the Permian rocks.

In general the coal-seams dip gently from the visible coalfield in the west to the concealed coalfield in the east. The history and development of the coalfield can be traced as it moved steadily eastwards from the outcrop and shallow workings on the eastern flank of the Pennines, in the period before 1850, to reach Doncaster by the early decades of the twentieth century.

Coals And Coal-Seams

The principal economic coal-seams are situated in the Middle Coal Measures; they are at their best with regard to number, thickness, quality and persistence. The most important of these is the Barnsley Bed Seam, which combines in the one seam, a soft bituminous house and coking coal, and a fine-quality hard steam-coal. The Silkstone, Middleton Main, Parkgate, Flockton and Haigh Moor seams are also important. In total, there are some thirty seams of workable thickness, the Barnsley Bed being the thickest of these. (A thin seam was regarded as one of 3ft or less in thickness).

In places ironstone (e.g. the Tankersley ironstone measures), fireclay and gannister occurred in close proximity to the coal-seams and were sometimes worked alongside them. The ironstone measures occurred in the Upper Coal Measures and were worked along the outcrop of the coal-seams from bell pits and shallow workings, in a similar manner to the early coal workings.

The coals of the Lower Coal Measures were usually only locally important in the visible part of the coalfield, usually along the outcrop. None of the Lower Coal Measure seams have been of great significance in South Yorkshire.

Coal-Seams In The Middle Coal Measures

The following coal-seams in the Middle Coal Measures, listed in descending order, have been worked in the South Yorkshire Coalfield.

Shafton (also known as **Billingley**) **Seam**
This seam had limited working in the South Yorkshire Area where it was worked at Dearne Valley and Goldthorpe Collieries and later at Kinsley and Ferrymoor/Riddings Drifts. Dearne Valley produced a coal suitable for steam-raising, manufacturing and brick-burning.

Swinton Pottery Seam
A poor quality coal, that was only worked on a small scale. It was worked at the outcrop around Swinton, where it was used for brick-burning.

Newhill Seam
At Manvers Main Colliery the seam had a thickness of approximately 5ft though generally split and of poor quality. The seam produced a domestic coal. The main outcrop of the seam is in the Wath and Swinton district, where it has been worked at the outcrop.

Meltonfield (also known as **Wath Wood**) **Seam**
This seam provided good, second-quality, house-coal and the small-coals were suitable for firing boilers. At Hickleton Main Colliery this coal was worked under the name Low Main Coal. At Wath Main and Houghton Main Collieries the small-coals from this seam were used for coke making. The seam has been opencast worked at a number of sites in South Yorkshire.

Winter Bed (also known as **Abdy**) **Seam**
A good free-burning, low-sulphur, domestic quality coal with a seam thickness of approximately 2ft 6in. It was worked at Manvers Main and Wombwell Main Collieries, as well as at opencast sites in South Yorkshire.

Beamshaw (also known as **Furnace**) **Coal**
This was one of the best house-coals in the area, possessing a low sulphur content. Slack from this coal was used for steam-raising and coke manufacture, and nuts were used for gas production.

Kent's Thick (also known as **High Hazel**) **Seam**
This seam provided a good house-coal, although the small-coal was not considered suitable for coke manufacture. The seam has also been opencast worked in South Yorkshire.

Barnsley (also known as **Warren House**) **and Barnsley Rider Seams**
One of the most famous coal seams in the world, the Barnsley Bed produced a steam-coal that surpassed even those of South Wales. Great colliery enterprises carved themselves vast royalties and steady profits out of this seam. In the early nineteenth century the miners of South Yorkshire enjoyed the reputation of being the best paid in the kingdom. In the 1930s stories of the colossal earnings of hewers in the Barnsley Bed were rife throughout the coalfields of Britain.

The Barnsley Bed was the most important coal-seam in South Yorkshire. It existed as a continuous, unbroken bed of coal over a total area in excess of 600 square miles, up to 900yds deep. This was the most widely worked seam in South Yorkshire, producing fifty per cent of the total output and providing direct employment for 60,000 men in the 1930s.

This seam was at its best in the vicinity of the town of Barnsley, from which it takes its name, and where it attained a seam thickness of 10-11ft of coal. Throughout South Yorkshire the seam thickness varied between 3ft and 11ft. To the east and south-east the seam thinned, but still maintained 6-8ft, with few washouts, over a large area. Beyond Rotherham, and to the south of Doncaster, thinning of the seam was more pronounced.

The seam had three main divisions – top, middle (or hards) and bottom. The Top: soft-coal was a band of bright coal – up to 5ft in thickness in the Barnsley area. The Hards: characteristic of the seam, was a band of dull hard coal, between 2ft 6in and 4ft 6in thickness, situated near to the base of the seam. This was further divided into an upper and lower portion, by a band of bright coal from 2 to 6in thickness. The Hards furnished a high-quality steam-coal. The Bottom: a bright coal, somewhat softer, generally known as the Bottom Softs.

Going north, commencing at Hickleton and Monckton Main Collieries, the seam split and the parting between the top soft-coal and the hard-coal varied between 5 and 45ft in thickness. Consequently only the lower portion of the seam, comprising the valuable hard-coal, was worked. As it approached the West Yorkshire Coalfield, the Barnsley Seam changed its character completely and the part of the seam that was worked was renamed the Warrenhouse Seam. South of Rotherham the nature of the seam also changed, becoming a non-coking steam-coal, renamed the Top Hard in Derbyshire and Nottinghamshire where it was extensively worked.

The seam furnished large-coal, used for steam-raising in locomotives, and for bunkering. Small-coal was used for coke manufacture, especially when blended (mixed) with Parkgate or Silkstone Seam coals, in the ratio 40:60. At Bentley, Brodsworth, Bullcroft, Markham, Rossington and Kiveton Park Collieries, the seam produced a non-coking coal, supplied to gasworks. Askern and Hatfield Collieries occasionally produced a gas-coal, though it was of inferior quality.

A thin seam, known as the Barnsley Rider, accompanied the Barnsley Bed in South Yorkshire, and parts of West Yorkshire.

Dunsil Seam

A seam varying in thickness from 3ft 6in to 5ft, split by dirt bands. Generally a poor-quality, 'dirty' coal, with high ash and sulphur content.

Swallow Wood (also known as **Haigh Moor**) **Seam**

There is often a great deal of confusion as to the name of this seam, but technically it should be called the Swallow Wood Seam in South Yorkshire, where it provided house-coal and gas-coal, and was regarded as a second-class steam-coal. At New Monckton Colliery the seam provided a very good gas-coal and a good house-coal, though at other collieries only a medium quality coal. The coal was mixed for coke making, while small-coal was used for firing boilers. To the north of Barnsley the seam was of better quality.

In some parts of South Yorkshire the Haigh Moor was a separate coal-seam, lying below the Swallow Wood, and worked as a separate coal-seam.

Lidgett Seam

This was a general house-coal in South Yorkshire. In some areas the seam provided manufacturing and steam-coal.

Flockton Seam

This seam provided a good house-coal and, with the exception of the coal produced by Waleswood Colliery, was mixed for use as a coking coal. It has also been opencast worked.

Fenton Seam

Worked in South Yorkshire and used chiefly for blending (mixing) with Parkgate Gas-coal as the Parkgate Seam was worked out. Small-coals were used for coke manufacture. In South Yorkshire the seam has also been opencast worked.

Parkgate Seam

Second in importance to the Barnsley Seam, its importance grew in the early part of the twentieth century as the Barnsley Seam was worked-out, this seam furnished a very fine gas-coal. It was worked on a wide scale in South Yorkshire and was at its best between Rotherham and Sheffield.

The seam comprised three divisions; Tops, Middle (Hards) and Bottoms. The upper Top and lower Bottom of the seam were inferior quality, producing high ash and being sulphurous in nature were not usually mined. The seam furnished industrial coals and was used extensively for gas manufacture. Coals produced by this seam were frequently blended with Barnsley Bed Coals. The Parkgate Seam, however, did not provide a good house-coal. Over a large area north of Barnsley the Parkgate Seam thinned and was known as the Old Hards. Despite this thinning, the seam was extensively worked.

Thorncliffe Seam

Widely worked in South Yorkshire where it provided a good quality house-coal and gas-coal. The seam furnished a low ash content coal, which was good for steam-raising. Small-coal was utilized for coke manufacture.

Swilley (or **New Hards**) **Seam**

This seam was only worked in the Dodworth-Barnsley area and provided gas and coking coals, and some steam-coal.

Silkstone Fourfoot (or **Wheatley Lime**) **Seam**

This seam was worked to a limited extent in the Barnsley area, providing coal suitable for household and industrial use. It was also worked opencast in South Yorkshire.

Silkstone Seam

Extensively worked over the coalfield, the seam furnished an excellent gas-coal for horizontal retorts and was particularly good, mixed with small-coal, for coke manufacture. The seam furnished a coal of great purity, providing a low-sulphur coal, and was the chief gas and coke manufacturing coal. The seam also provided a good house-coal. The seam was at its best over the central region of the coalfield, deteriorating to the north and south, due to the intrusion of dirt partings. Taking its name from the village where the seam outcropped, it was well known in the coal trade. The seam achieved national fame following the 1851 London Exhibition, when a specially mounted 3cwt block of Silkstone coal was exhibited. The Silkstone Seam was the deepest seam in the Middle Coal Measures.

Coal-Seams In The Lower Coal Measures

The following coal-seams have been worked in the South Yorkshire Coalfield in the Lower Coal Measures.

Whinmoor Seam
This seam provided a poor quality gas and house-coal and was principally used as an industrial coal. It has been opencast worked in the South Yorkshire area.

Halifax Hard (also known as **Gannister**) **Seam**
The coal from this seam was used principally in local manufacturing industry and for steam-raising. It was only worked in conjunction with the accompanying gannister and associated fireclays. Worked along the outcrop, the seam was extensively mined between Sheffield and Penistone.

Halifax Soft or Coking Seam
This seam was not extensively worked, but was principally worked by drift mines along the outcrop between Sheffield and Stocksbridge. The seam furnished a gas-coal but was principally used as a coking coal. At Stocksbridge the seam yielded a good coking coal of exceptionally low-sulphur content.

Pot Clay Coal-seam
This seam was worked on a small scale in the South Yorkshire area. The coal was used for pipe manufacture.

3
South Yorkshire Colliery Companies

There were two principal groups of colliery owners in South Yorkshire. The first of these was closely allied with the iron and steel trades, particularly of the Sheffield district, with collieries located in the southern part of the coalfield. These comprised local companies such as: John Brown & Co.; the United Steel Co.; the Staveley Coal & Iron Co. and the Sheepbridge Coal & Iron Co. The ranks of these companies were swollen by steel companies from further afield, such as: Dorman Long, from the Tees; Pease & Partners, from County Durham, and the Barrow Haematite Iron & Steel Co., from Cumbria. This was a group of large producers whose interests were essentially the manufacture of steel, to whom the collieries were an ancillary to steel production. These companies were highly integrated, with businesses ranging from the mining of coal, limestone and ironstone, to steel manufacture and engineering. By the mid-1930s the Staveley and Sheepbridge interests controlled the great mass of coal production in the twenty-five mile stretch between Chesterfield, in Derbyshire, and Doncaster. Between them the two concerns owned twenty collieries, with a capacity in excess of 14m tons, and a workforce of some 38,000.

The second group of colliery owners principally worked the collieries in the central region of the coalfield, to the east of Barnsley. These collieries generally confined their activities to coal and its ancillary industries, such as coke and brick-making. They were represented by Carlton Main and the South Kirkby Colliery groups.

The Mitchelson Group

Sir Archibald Mitchelson, a close friend of Lord Rhondda, the creator of the Cambrian Combine in South Wales, brought the best in training to the companies over which he held directorates. In 1922-1923, under the ownership of the Old Silkstone Collieries Ltd, he brought together Goldthorpe Collieries Ltd, the Allerdale Coal Co. (Cumbria), Garforth Collieries (West Yorkshire), Old Silkstone Chemical Works Ltd and the Dodworth Estate & Finance Co. The group was well-planned and owned one of the most famous coke-oven and by-product plants in the country at Barugh, together with the neighbouring Old Silkstone Chemical Works.

The Sutherland Group

Sir William Sutherland married Anne Fountain in 1921 and moved from a political career to become a colliery owner. In the mid-1930s the Sutherland group of collieries, located to the north of Barnsley, comprised Wharncliffe Woodmoor Colliery Co., Fountain & Burnley Ltd, George Fountain & Son Ltd and Fountain & Son Ltd. In all, some seven collieries, employing about 7,300 men, owned entirely by Sir William and Lady Anne Sutherland.

Amalgamated Denaby Collieries Ltd (see Denaby & Cadeby Main Colliery Co.)

The Denaby Main Colliery Co. was registered in March 1868, with an authorized capital of £110,400. In 1893, following the opening out of Cadeby Main Colliery, the company was renamed the Denaby & Cadeby Main Collieries Co. Ltd. In March 1927 the company was amalgamated with a number of other South Yorkshire collieries, and formed into the Yorkshire Amalgamated Collieries Co. Ltd, which had a capital of £3.2million. It comprised Rossington Main, Dinnington Main, Maltby Main, Darton Main and the Strafford Colliery Co. Ltd. On 27 April 1936 the company was incorporated under the title the Amalgamated Denaby Collieries Co. Ltd, a title retained until the formation of the NCB, on 1 January 1947. In the early 1940s the company employed some 10,000 and produced an annual output of 4m tons.

Askern Coal & Iron Co. Ltd

In March 1910 a private company was set up and registered, with a capital of £400,000, to sink and operate a colliery at Askern, near Doncaster. The Askern Coal & Iron Co. was set up primarily to acquire and develop the extensive mining options at Askern, which had been granted to the Don Coal & Iron Co. Ltd, the initial take of which comprised 7,000 acres with a further 7,000 acres available. The colliery was owned by the Askern Coal & Iron Co. until nationalization. The Bestwood Coal & Iron Co., of Nottinghamshire, and the Blaina Colliery Co., of Monmouthshire, took a leading role in setting up the new company.

Barber, Walker & Co. Ltd

Messrs Barber, Walker, of Eastwood in Nottinghamshire, was a long established mining concern, tracing its origins back to c.1680. The company, like many of its mining contemporaries in the Nottinghamshire Coalfield, was concerned about the gradual exhaustion of the Top Hard Seam (the equivalent of the Barnsley Seam in the Nottinghamshire Coalfield) and were willing to go further afield in search of their coal workings. To this end they sunk Bentley Colliery in South Yorkshire.

Barrow Barnsley Main Collieries Ltd

Barrow Colliery was sunk by the Barrow Haematite Iron & Steel Co., which produced iron and steel in the Furness region of north Lancashire (now part of Cumbria). Barnsley Main Colliery Co. was registered on 15 May 1899. Barrow Colliery amalgamated with Barnsley Main and Monk Bretton Collieries in 1932, forming Barrow Barnsley Main Collieries Ltd, which retained ownership of the collieries until nationalization. In the early 1940s the company produced an annual output of 1.3m tons and employed 4,210.

Brodsworth Main Colliery Co. Ltd

The Brodsworth Main Colliery Co., a joint venture between the Hickleton Main Colliery Co. and the Staveley Coal & Iron Co., was formed and registered in 1905. The new company had a capital of £300,000. The Staveley company came to have a large controlling influence in the Doncaster district in the early years of the twentieth century.

John Brown & Co. Ltd

John Brown & Co., of the Atlas Steel & Iron Works, Sheffield, were one of the best known steel-making concerns in Sheffield, and well known in marine engineering, munitions and ship-building. They were also colliery owners, with large concerns in the vicinity of Rotherham, including Rotherham Main and the nearby Aldwarke Main and Carr House Collieries. The company's colliery concerns were an important part of their operation. In 1899 the company employed 3,500 at their Atlas Steel Works and 4,500 at their collieries. In the first decade of the twentieth century the company had an output of 2.5m tons of coal per annum.

Bullcroft Main Colliery Co. Ltd

The Bullcroft Main Colliery Co. Ltd was registered in April 1908.

Carlton Main Colliery Co.

The creator of the Carlton Main Colliery Co. was James Jenkin Addy. Addy's father began the sinking of the old Carlton Colliery in 1872, with a capital of £73,000, to work the Barnsley Bed Seam at Carlton, on the outskirts of Barnsley. Under James Addy's leadership, the company expanded steadily, and in 1896, acquired the Grimethorpe Colliery Co. Ltd. In 1900 the two companies were united under the name the Carlton Main Colliery Co. Ltd. In 1903 the company commenced the sinking of Frickley Colliery. Carlton Colliery was almost exhausted in the early years of the twentieth century and was abandoned in 1909. Wartime profits fuelled further expansion, and in 1919, the company acquired the adjacent Hodroyd Coal Co. Ltd, which owned two small collieries at Brierley and Ferrymoor. In 1920 South Elmsall Colliery was sunk within the grounds of Frickley colliery to work the Shafton Seam. In 1926 the company acquired Llay Main Collieries, near Wrexham, North Wales, and Hatfield Colliery, Doncaster, in 1927. In 1919 the company took a strong financial interest in Rea Ltd, a Liverpool based coal-handling company with shipping facilities on the Mersey. In addition, the company formed the Carlton Collieries Association, an organization acting as a selling agent, similar to the Doncaster Collieries Association. By the late 1920s the company had a production capacity of some 6m tons per annum and employed a workforce of approximately 15,000. The colliery group ranked amongst the most profitable in South Yorkshire, even making a profit during the depression of the 1920s.

Cortonwood Colliery Co. Ltd

The Cortonwood Colliery Co. was formed in 1872, having previously been called the Brampton Colliery Co. It was formed specifically to exploit the Barnsley Seam on the Elsecar Branch of the Dearne & Dove Canal. Cortonwood Colliery was sunk between 1873 and 1875 by the company, who retained ownership of the colliery until nationalization. On 22 October 1883 a Memorandum of Association of the Cortonwood Colliery Co. Ltd was drawn up, with a total capital of £175,000.

Dalton Main Collieries Ltd

In 1898 the Roundwood & Dalton Colliery Co. was formed with a capital of £200,000. The new company acquired Roundwood Colliery, a long-established colliery in the district, and a large tract of land at Silverwood, to sink a new colliery and build a railway. In December 1899 the Dalton Main Colliery Co. was formed, and in 1900 sinking of Silverwood Colliery started.

Darton Main Colliery Co. Ltd

This colliery company was registered on 13 August 1913.

Dearne Valley Colliery Co. Ltd

The Dearne Valley Colliery Co. was formed in 1901, with a capital of £30,000, to sink and work the Dearne Valley drift mine.

Denaby & Cadeby Main Colliery Co.

Denaby Colliery was a high-risk venture undertaken by the established West Yorkshire coal owners Messrs Pope & Pearson. In March 1868 the Denaby Main Colliery Co. was registered with a share capital of £110,400. In the early 1890s Cadeby Main, a sister colliery, was sunk a mile or so further up the River Don valley. In 1893 a new company, the Denaby & Cadeby Main Colliery Co. Ltd was formed. In March 1927 they were amalgamated into the Yorkshire Amalgamate Collieries Ltd.

Dinnington Main Coal Co. Ltd

The Sheffield Coal Co. and Sheepbridge Coal & Iron Co. entered into partnership, and the resultant Dinnington Colliery Co. was registered on 1 November 1900, with a capital of £187,500.

Doncaster Amalgamated Collieries Ltd

This company was formed on 9 February 1937 by the Staveley Group, with an authorized capital of £7.75million. The company comprised Brodsworth Main, Bullcroft Main, Hickleton Main, Markham Main, Yorkshire Main and Firbeck Main (Nottinghamshire) Collieries. The company also incorporated the Doncaster Collieries Association Ltd, which was purely a buying and selling organization, previously set up by the Staveley Group.

Earl Fitzwilliam's Collieries Co.

In the nineteenth century the Earl operated a number of small collieries in the vicinity of Elsecar, in South Yorkshire. In the early twentieth century these were exhausted or nearing exhaustion, leading the Earl to sink two collieries, Elsecar and New Stubbin, on his land in South Yorkshire, initially to work the deeper Parkgate Seam, the Barnsley Seam having been worked out over the area worked by the two collieries. On 7 February 1933 the Earl Fitzwilliam Colliery Co. Ltd was registered with a capital of £1.5 million. In the early 1940s the Earl's collieries produced 1.2m tons of coal per annum.

Fountain & Burnley Ltd

Fountain and Burley Ltd owned North Gawber Colliery from 1882 until nationalization. Between 1910 and 1912 they sank two shafts at the site, which became known as Woolley Colliery, owning this colliery until nationalization.

Fox, Samuel & Co. Ltd

Stocksbridge Colliery was developed by Samuel Fox, a steel-making company of Stocksbridge, and a subsidiary of the United Steel Co.

Goldthorpe Collieries Ltd

Goldthorpe Collieries Ltd was registered on 3 September 1923 with a capital of £50,000. The shares owned by the Old Silkstone Colliery Co. who purchased Goldthorpe Colliery.

Hatfield Main Colliery Co. Ltd

In December 1910 the Hatfield Main Colliery Co. Ltd was registered with a capital of £300,000. Between 1911 and 1917 the company sunk Hatfield Main Colliery. In 1927 Hatfield Main Colliery was acquired by the Carlton Main Colliery Co.

Hickleton Main Colliery Co. Ltd
Hickleton Main Colliery Co. was formed in the 1890s. In 1937 the company joined the newly formed Doncaster Amalgamated Collieries Ltd, who owned the colliery until nationalization.

Highgate Colliery Co. Ltd
Sinking of Highgate Colliery was commenced in 1916, by the Highgate Colliery Co. who retained ownership until nationalization.

Hodroyd Colliery Co. Ltd
The company owned Brierley and Ferrymoor Collieries. In 1919 the company was taken over by the Carlton Main Colliery Co.

Houghton Main Colliery Co. Ltd
Houghton Main Colliery was sunk by the Houghton Main Colliery Co., who operated the mine until nationalization. In 1875 the company was registered with a capital of £100,000. At nationalization, on 1 January 1947, the colliery had a capital value of £645,021.

Kiveton Park Coal Co. Ltd
In March 1864 the Kiveton Park Colliery Co. formed with an estimated capital of £200,000. In 1896 the company was incorporated as a limited liability company.

Maltby Main Colliery Co. Ltd
The Maltby Main Colliery Co. was registered in 1906 with a capital of £350,000, of which the Sheepbridge Coal & Iron Co. was the major shareholder, holding £200,000. In 1907 the sinking of Maltby Main Colliery began. In March 1927 the company joined with other South Yorkshire collieries forming the Yorkshire Amalgamated Collieries Co. Ltd.

The Manvers Main Colliery Co. Ltd
This company was registered on 21 February 1899 with a capital of £1,000,000. The company negotiatied with the United Steel Co. for the acquisition of Kilnhurst and took over the management of the colliery from 1 February 1945. The company also owned Barnburgh Main Collieries.

Markham Main Colliery Co. Ltd
In June 1913 Sir Arthur Markham secured the lease to the mineral rights at Armthorpe, near Doncaster, from Earl Fitzwilliam. In 1916 sinking began at Armthorpe. Sir Arthur died soon after sinking commenced and the colliery became known as Markham Main in his honour.

Mitchell Main Colliery Co. Ltd
In 1883 the Mitchell Main Colliery Co. was registered with a capital of £32,000. The company owned Mitchell Main and, at various times, Darfield Main. In 1925 the Mitchell interests were sold.

New Monckton Collieries Ltd
The Monckton Main Coal Co. Ltd was registered as a limited liability company on 16 September 1874, with a capital of £120,000. In 1901 a reconstruction of the undertaking took

place. The old Monckton Coal Co. Ltd was wound up and a new company, called New Monckton Collieries Ltd, was formed with a transferred capital of £220,000. New Monckton Collieries was incorporated on 25 July, taking over the business and assets of the old company. The new company had a capital of £500,000, comprising Monckton Main and Hodroyd Collieries, 180 beehive coke ovens, by-product plant and a brickyard. New Monckton Collieries Ltd was floated as a public company, and the first shares, allocated to the public, were approved at a meeting in August 1901.

Newton, Chambers & Co. Ltd, Thorncliffe Collieries

In 1793 George Newton and Thomas Chambers, partners in the Phoenix Foundry in Sheffield, and Henry Longden, signed a twenty-one year lease from Earl Fitzwilliam for land at Thorncliffe. The land was to build an ironworks, and to mine coal and ironstone. In total Newton Chambers leased 10,139 acres, of which 5,700 acres were from Earl Fitzwilliam and 1,350 acres from the Duke of Norfolk. Newton, Chambers & Co. Ltd were registered in 1881 with a capital of £650,000. The company's Rockingham, Thorncliffe, Smithywood and Grange Collieries worked an area of fifteen square miles of coal between Barnsley and Rotherham. By the early 1930s eighty per cent of their output was machine won. Newton, Chambers & Co. produced 1,083,935tons of coal from these four collieries in the year ending 31 December 1935. The company mined ironstone, for their ironworks at Thorncliffe, until 1880, and coal until nationalization.

Nunnery Colliery Co.

The Nunnery Colliery Co. was registered on 28 September 1874 with a capital of £60,000, when a group of Sheffield capitalists acquired the lease for Nunnery colliery, subsequently working it under the title the Nunnery Colliery Co. In its formative years the company had a nominal capital of £90,000. The company sank Handsworth Colliery between 1901-1903. Both collieries were owned and run by the company until nationalization in 1947. In the early 1940s the company employed 1,630 and had an output of 800,000 tons.

Old Silkstone Collieries Ltd

The Old Silkstone and Dodworth Coal & Iron Co. Ltd was founded in 1862. In 1899 a new company the Old Silkstone Collieries Ltd was founded. The company owned Dodworth, Stanhope Silkstone and Silkstone Fall Collieries and Higham pumping station.

Pease & Partners Ltd

Pease & Partners were a well-established coal and steel company of Darlington. They sank Thorne Colliery, between 1909 and 1926, to provide coking coal for their iron and steel-making plants. The difficulty of the sinking and the protracted time-scale put a severe financial strain on the company.

Rossington Main Colliery Co. Ltd

Owned by the Sheepbridge Coal & Iron Co. and John Brown & Co., the Rossington Main Colliery Co. was registered on 15 July 1911 with a capital of £750,000. The company began the sinking of Rossington Colliery in 1912.

Rothervale Collieries Ltd

In 1842 a small colliery was started at Fence, on the outskirts of Sheffield. In 1862 Fence Colliery Co. was formed. It acquired Orgreave Colliery in 1870 and, a few years later, developed Treeton Colliery. In 1874 the Fence Colliery Co. was wound up and reformed as the Rothervale Collieries Ltd., with a nominal capital of £300,000 and 2,865 acres of coal reserves. In 1918 the United Steel Co. was registered and in the same year Rothervale Collieries became a branch of the new company. The United Steel Co. Ltd comprised Steel, Peech & Tozer, Samuel Fox Ltd, the Workington Iron & Steel Co. Ltd and the Frodingham Iron & Steel Co. Ltd. In 1918 the construction of the Orgreave Coking Plant began. One of the aims of the United Steel Co. was to establish assured markets for coal and coke from their new Orgreave plant.

Sheffield Coal Co. Ltd

The Sheffield Coal Co. was formed in 1805, and was the oldest colliery company in the district. Though the company traded as the Sheffield Coal Co., its correct name was Jeffcock, Dunn & Co. By 1898 the company employed 2,500-2,600. In 1927 the Sheffield Coal Co. and the Aston Coal Co. amalgamated. In late 1927, or early 1928, T.W. Ward of Sheffield acquired a considerable interest in the company. Dinnington Colliery was originally projected by the Sheffield Coal Co. who realized that they did not have the financial resources to sink the colliery, and entered into partnership with the Sheepbridge Coal & Iron Co. The resultant Dinnington Colliery Co. was registered on 1 November 1900 with a capital of £187,500. In June 1937 the company was acquired by the United Steel Co. Ltd.

Skinner & Holford Ltd, Waleswood Collieries

Messrs Skinner & Holford, the proprietors of Waleswood Colliery, had leased considerable areas of mineral property on the Yorkshire-Derbyshire border and sunk Waleswood Colliery. The company of Skinner & Holford Ltd. was registered in 1884 with a capital of £100,000.

South Kirkby, Featherstone & Hemsworth Collieries Ltd

In 1879, the newly sunk South Kirkby Colliery looked as though it was destined to fail, when the South Kirkby Colliery Co. was formed. It was registered on 24 January 1882 with a capital of £100,000. In 1905 Hemsworth Colliery and the South Kirkby were acquired. Featherstone and Hemsworth Collieries Co. was registered in September 1906 with a capital of £900,000. The collieries in the group were owned by the new company until nationalization.

Sheepbridge Coal & Iron Co.

With an interest in the iron industry and the utilization of its by-products, this company, like the Staveley Company, arose from the exploitation of the shallow coal and ironstone deposits of North Derbyshire. As the ironstone deposits of North Derbyshire were exhausted, the company moved to new fields in the East Midlands and Yorkshire. Each company was the nucleus for a mass of colliery subsidiaries. Interestingly, there was no rivalry between the two companies, and they shared many common interests. The Sheepbridge company promoted the Dinnington Main Colliery Co. in 1900, and the Maltby Company in 1907. In 1911, in association with John Brown & Co., the Sheepbridge

Company formed the Rossington Main Colliery Co. Ltd. In 1927 the Sheepbridge company sub-sidiaries, together with the Denaby & Cadeby Main Colliery Co., were grouped under the title the Yorkshire Amalgamated Collieries Ltd.

Staveley Coal & Iron Co.

With a background similar to the Sheepbridge Company, Staveley developed by exploiting the shallow coal and ironstone deposits of North Derbyshire, and had an interest in the iron industry and the utilization of its by-products. Like the Sheepbridge company, the exhaustion of the North Derbyshire ironstone deposits forced a move to the new fields in the East Midlands and Yorkshire. The company was in the forefront of the rush to develop the Doncaster area of the South Yorkshire Coalfield. In 1905 the Brodsworth Main Colliery Co. was promoted, in conjunction with the Hickleton Main Colliery Co., another Staveley company. In 1907 the Yorkshire Main Colliery Co. was formed, followed by the Markham Main Colliery Co., which was formed in conjunction with the Bullcroft Main Colliery Co., another of the company's subsidiaries. South Yorkshire collieries owned by the company included Brodsworth, Bullcroft, Hickleton, Markham and Yorkshire Main. There was a considerable market for the coal produced by the group and the company had a sound financial record.

Stewarts & Lloyds Ltd

Allied to the United Steel Co. by a co-operative agreement, the important Stewarts and Lloyds company had subsidiaries in the iron trade on the Clyde, in the Midlands and abroad. With links to coal producers in the Clyde valley, the company was the owner of Kilnhurst Colliery.

Tinsley Park Colliery Co. Ltd

The Tinsley Park Colliery Co. was registered in 1898, though the Tinsley Park Colliery had started work in the early nineteenth century. In 1936 the company acquired Kilnhurst Colliery and, in 1939, Messrs J. & G. Wells Ltd.

United Steel Co. Ltd

In 1918 the United Steel Co. was registered, and in the same year Rothervale Collieries became a branch of the new company. The United Steel Co. Ltd comprised Steel, Peech & Tozer, Samuel Fox Ltd, the Workington Iron & Steel Co. Ltd, and the Frodingham Iron & Steel Co. Ltd. In 1918 the construction of the Orgreave Coking Plant began. One of the aims of the United Steel Co. was to establish assured markets for coal and coke from their new Orgreave plant. By the mid-1930s the company produced over 2m tons of coal per annum. The United Steel Co. took over the Tinsley Park Colliery Co. in December 1944.

Upton Colliery Co. Ltd

In 1920 the Co-operative Wholesale Society of Manchester secured the mining rights to some 4,000 acres at Upton, but did not proceed with the sinking of a colliery. In November 1923 a syndicate, formed by the Cortonwood Colliery Co. and Bolckow Vaughan & Co., a steel-making company from Teeside in North Yorkshire, registered the Upton Colliery Co. with a capital of £600,000. By this time some 8,000 acres had been leased. In November 1929 Messrs

Dorman Long took over Bolckow Vaughan. The Cortonwood Colliery Co. sold their share to Dorman Long, and in 1939, the original company was liquidated and absorbed into Dorman Long & Co. Ltd.

Wath Main Colliery Co. Ltd

The Wath Main Colliery Co. was registered on 25 May 1900, with a capital of £250,000. The company owned and ran Wath Main Colliery until nationalization.

Wentworth Silkstone Collieries Ltd

A colliery was sunk at Wentworth Silkstone by one Samuel John Cooper, reaching coal in December 1857. The colliery appears to have been closed between 1887 and 1912, when the Wentworth Silkstone Colliery Co. was formed, which ran the colliery until nationalization.

Wharncliffe Silkstone Colliery Co. Ltd

The Colliery was sunk in 1853-1854. In 1879 the Wharncliffe Silkstone Colliery Co. was formed to run the business. The colliery took its name from the chief royalty owner, Lord Wharncliffe, and the Silkstone Seam, which was the first seam to be worked. The Wharncliffe Silkstone Colliery Co. owned and worked the colliery until nationalization.

Wharncliffe Woodmoor Colliery Co. Ltd

In September 1870 some 350 acres of the Woodmoor and Winter Seams at Carlton, under Lord Wharncliffe's estate, were leased to Joshua Willey. The colliery planned by Willey, New Willey Colliery, was to be quite small. Willey sank two shafts, reaching the Woodmoor Seam in November 1871. Soon afterwards the colliery was sold, and in 1873 a new colliery company, the Wharncliffe Woodmoor Coal Co., was formed. The effect of the collapse of the coal boom in the mid-1870s was felt at the colliery, and in November 1876 the colliery was put up for sale, to be sold in one lot, as a going concern. In 1881 Howard Allport bought the colliery and became the owner of Wharncliffe Woodmoor. In 1883 Allport converted his privately owned colliery into a limited liability concern, which became the Wharncliffe Woodmoor Colliery Co. Ltd, registered with a capital of £30,000.

Wombwell Main Colliery Co. Ltd

Wombwell Main Colliery was sunk between 1853 and 1855. The colliery subsequently converted into a private limited company, under the name the Wombwell Main Co. Ltd. The Wombwell Main Co. was to remain the owner of the colliery until nationalization.

Yorkshire Amalgamated Collieries

Yorkshire Amalgamated Collieries Ltd was registered in 1927, to take control of the Denaby & Cadeby Main Collieries Co., Dinnington Main Coal Co., Maltby Main Colliery Co. and Rossington Main Colliery Co. The new company controlled the mineral rights under some fifty square miles of the South Yorkshire Coalfield. It had a designed output capacity of 6.75m tons of coal per annum from a workforce of over 13,000. The company operated more than 10,000 railway wagons from a central control office, and despatched some forty loaded coal trains (with a combined total length of over fifty-two miles) each day. Closely allied with the company was Messrs William France Fenwick, who owned a fleet of steamers specially constructed for the carriage of coal around the coast and to the continent.

Yorkshire & Derbyshire Coal & Iron Co. Ltd

Founded by James Addy and William Wake, a Sheffield solicitor, sometime around the middle of the nineteenth century, the company sank a colliery called Springfield Main at Cold Aston and Dronfield near Sheffield, which they were working in 1862. The Carlton Main Colliery Co. Ltd had its genesis in the Yorkshire & Derbyshire Coal & Iron Co. Ltd, and was formed in 1872 to work the Barnsley Bed Seam at Carlton, on the outskirts of Barnsley.

Yorkshire Main Colliery Co.

The Staveley Coal & Iron Co. of Derbyshire, were interested in the colliery sinkings being undertaken in the Doncaster area of the South Yorkshire Coalfield. In 1907 it acquired mineral rights in the vicinity of the village of Edlington. In 1909 the sinking of a colliery began. Initially known as Edlington Main, in September 1909 the name was changed to Yorkshire Main and in 1913 the Yorkshire Main Colliery Co. was formed.

4
The Coal Industry Before Nationalization

For more than 150 years, including the first decade after nationalization, coal played a dominant role in the industrial growth and economic expansion of Britain. It 'fired' the industrial revolution and established and maintained the lifelines of a vast empire. For more than a century coal had sped the Royal Navy around the globe and filled the bunkers of a huge merchant navy. It was the foundation of British overseas trade and formed an important export commodity in its own right.

The death knell for the industry came before the First World War, with the introduction of ships powered by oil, and with the relentless shrinking of the coal-export trade as emerging nations developed their indigenous coal industries. Between 1913, the peak year for production, and 1953, investment in the coal industry had been insufficient to offset the inevitable decline in capacity.

The 1930 Coal Mines Act established the Coal Mines Reorganization Committee. In an attempt to overcome some of the problems of an industry beset with fragmentation, the committee aimed to encourage the amalgamation of mining companies, with a view to increasing their efficiency and productivity. As a result of the Act, a number of amalgamations took place in South Yorkshire.

Until the 1938 Coal Act, ownership of coal deposits remained with the proprietor of the land, who charged royalties to the mine operators. The 1938 Act transferred ownership of coal deposits to the state and provided for payment of compensation to the landowners.

The post-war Labour Government passed the 1946 Coal Industry Nationalization Act, which transferred the coal industry and its assets into state ownership, on payment of compensation to the owners. Nationalization of the railway, electricity and gas industries was to follow quickly. The primary purpose of the National Coal Board (NCB), as defined in the Act of 1946, was:

> *To work the coal, to secure the efficient organization of the industry and to make available supplies of coal of such qualities and sizes, in such quantities and at such prices, as seem to the Board best calculated to further the public interest.*

5
The Nationalized Coal Industry

On 1 January 1947 the coal industry was nationalized. A notice was posted at every colliery in Britain, which read: 'This colliery is now managed by the NATIONAL COAL BOARD on behalf of the People. January 1, 1947.' It was not until 1 April 1952 that opencast mining, which had been started as an emergency measure by the Ministry of Fuel and Power in 1942, to boost production, was taken over by the NCB.

Following the nationalization of the British coal industry, the NCB set about the task of modernizing and reorganizing the coal industry which they had inherited. It paid compensation of £164,660,000 to the original owners for the main assets of the collieries. It also took over the outstanding compensation to former royalty owners, which at the time was valued at £78,457,008, together with other compensation for subsidiary assets, such as brickworks.

The NCB had taken on a massive undertaking comprising some 958 collieries, the property of some 800 concerns, more than 400 small mines, fifty-five coke ovens (producing coke, tar, benzol and sulphuric acid), twenty manufactured fuel and briquette plants, eighty-five brick

Notice posted at nationalized collieries on 1 January 1947.

and pipe works, colliery power stations and waterworks, aerial ropeways and railway sidings and other supporting infrastructure. In addition the Board had the management of 225,000 acres of farmland, 141,000 houses (many of which were drab nineteenth century back-to-back houses) and buildings ranging from offices, shops, and hotels, to swimming baths, a cinema and a slaughterhouse. In order to carry the 28m tons of coal consumed by the electricity generation industry alone, the NCB required a stock of 200,000 wagons.

To run this enterprise the NCB had inherited a workforce of 796,000, comprising workmen, managers, administrative and clerical staff, some four per cent of the total workforce of Britain. The new enterprise had an annual turnover of £370million, and planned to mine 20,000m tons of coal during the next hundred years. Overall output per man-shift stood at just over one ton and 2.4 per cent of output was mechanized. The output of the coal industry in 1947 stood at 184m tons, together with a further 10m tons produced by opencast sites.

At Nationalization the 'average' colliery produced an output of 245,000 tons per annum and worked seams varying from 20 to 200in thick. Despite this one-third of all collieries produced less than 100,000 tons, while fifty collieries produced 700,000 tons plus. Of the fifty-seven or so collieries which comprised the South Yorkshire Coalfield, sixteen produced less than 245,000 tons, twenty-one produced 245,000 to 500,000 tons, ten produced 500,000 to 750,000 tons, seven produced 750,000 to 1m tons, and three collieries produced over 1m tons. South Yorkshire collieries were in the big league of producers.

It was decided that the newly formed NCB would have some fifty Areas, each having an output of about 4m tons. The Areas were grouped into eight Divisions, each Division corresponding to a major coalfield. The Divisions were: No.1, Scottish Division – five areas and 206 collieries, covering the Scottish Coalfield; No.2, Northern Division – ten areas and 222 collieries, covering the Durham, Northumberland, Cumberland and Westmorland Coalfields and a small area of the North Riding of Yorkshire; No.3, North-Eastern Division – eight areas, and 118 collieries, covering the Yorkshire Coalfield, except for the small area of North Yorkshire covered by the Northern Division collieries; No.4, North-Western Division – seventy-four collieries and five areas, covering the Lancashire and North Wales Coalfields; No.5, East Midlands Division – 107 collieries and eight areas, covering the Derbyshire, Nottinghamshire, South Derbyshire and Leicestershire Coalfields; No.6, the West Midlands Division – four areas and sixty-seven areas, covering the North Staffordshire, Cannock Chase, South Staffordshire, Shropshire and Warwickshire Coalfields; No.7, the South-Western Division – with 230 collieries and seven areas, covering the South Wales, Forest of Dean, Bristol and Somerset Coalfields; finally, No.8, the smallest Division – one area and four collieries, covered the detached Kent Coalfield.

The Yorkshire Coalfield, which comprised the distinct areas of West and South Yorkshire, was formed into the No.3, or North-Eastern, Division, of the newly nationalized coal industry. This Division was divided into eight areas, each sub-divided into 'A' and 'B' sub-areas, with the exception of the No.8 Castleford Area, which was sub-divided into 'A', 'B' and 'C' areas.

The eight Divisions of the National Coal Board, 1 January 1947.

No.1, Worksop Area

Sub-Area 'A' (seven collieries)
- Beighton (Brookhouse) Colliery
- Harworth Main Colliery*
- Maltby Main Colliery
- Nunnery Colliery
- Orgreave Colliery
- Thurcroft Colliery
- Treeton Colliery.

Sub-Area 'B' (five collieries)
- Dinnington Main Colliery
- Firbeck Main Colliery
- Kiveton Park Colliery
- Manton Colliery*
- Waleswood Colliery

(* These collieries were situated in Nottinghamshire.)

No.2, Doncaster Area

Sub-Area 'A' (five collieries)
- Askern Main Colliery
- Brodsworth Main Colliery
- Bullcroft Main Colliery
- Hickleton Main Colliery
- Rossington Colliery
- Yorkshire Main Colliery

Sub-Area 'B' (seven collieries)
- Bentley Colliery
- Goldthorpe Colliery
- Hatfield Colliery
- Highgate Colliery
- Markham Main Colliery
- Thorne Colliery.

No.3, Rotherham Area

Sub-Area 'A' (six collieries)
- Aldwarke Colliery
- Cortonwood Colliery
- Elsecar Main Colliery
- New Stubbin Colliery
- Rotherham Main Colliery
- Silverwood Colliery.

Sub-Area 'B' (six collieries)
- Barnburgh Main Colliery
- Cadeby Main Colliery
- Denaby Main Colliery
- Kilnhurst Colliery
- Manvers Main Colliery
- Wath Main Colliery

No.4, Carlton Area

Sub-Area 'A' (six collieries)
- Frickley Colliery
- Hemsworth Colliery
- Monckton Nos 1, 2 and 3 Colliery
- South Elmsall Colliery
- South Kirkby Colliery
- Upton Colliery.

Sub-Area 'B' (seven collieries)
- Brierley Colliery
- Darfield Main Colliery,
- Dearne Valley Colliery
- Ferrymoor Colliery,
- Grimethorpe Colliery
- Houghton Main Colliery
- Mitchell's Main Colliery,

No.5, South Barnsley Area

Sub-Area 'A' (six collieries)
- Barnsley Main Colliery
- Barrow Colliery
- Dodworth Colliery
- Monk Bretton Colliery
- Silkstone Common Colliery **
- Wentworth Silkstone Colliery.

Sub-Area 'B' (eleven collieries)
- Bullhouse Colliery
- Grange Colliery
- Hazlehead Colliery **
- Rockingham Colliery
- Sledbrook Colliery **
- Smithywood Colliery
- Stocksbridge Colliery
- Thorncliffe Colliery
- Wharncliffe Chase Colliery
- Wharncliffe Silkstone Colliery
- Wombwell Main Colliery

(** These were tiny collieries working the outcrop and producing between 5,000 and 15,000 tons per annum, employing a workforce of between forty and eighty.)

No.6, North Barnsley Area

Sub-Area 'A' (thirteen collieries)

Bullcliffe Colliery
Caphouse Colliery
Crigglestone Colliery
Denby Grange Colliery
Emley Moor Colliery
Grange Ash Colliery
Gregory Springs Colliery
Hartley Bank Colliery
Howroyd Colliery
Lepton Edge Colliery
Newmillerdam Colliery
Parkmill Colliery
Shuttle Eye Colliery.

Sub-Area 'B' (six collieries)

Darton Colliery
Haigh Colliery
North Gawber Colliery
Wharncliffe Woodmoor 1 2 & 3 Colliery
Wharncliffe Woodmoor 4 & 5 Colliery
Woolley Colliery

No.7, Wakefield Area

Sub-Area 'A' (eight collieries)

Birkenshaw Colliery
Manor Colliery
New Gomersal Colliery
Norwood Green Colliery
Old Roundwood Colliery
Shawcross Colliery
Thornhill Colliery
Toftshaw Moor Colliery

Sub-Area 'B' (nine collieries)

East Ardsley Colliery
Fanny Colliery
Lofthouse Colliery
Middleton Broom Colliery
New Sharleston Colliery
Nostell Colliery
Park Hill Colliery
St. John's Colliery
Sharleston West Colliery

No.8, Castleford Area

Sub-Area 'A' (eight collieries)

Allerton Bywater Colliery
Allerton Primrose Colliery
Allerton Silkstone Colliery
Allerton Victoria Colliery
Fryston Colliery
Peckfield Colliery (Micklefield)
Waterloo Main Colliery (Temple & Park)
Wheldale Colliery.

Sub-Area 'B' (five collieries)

Newmarket Silkstone Colliery,
Snydale Colliery
Water Haigh Colliery,
West Riding & Silkstone Colliery,
Whitwood & Savile Colliery

Sub-Area 'C' (three collieries)

Ackton Hall Colliery
Glass Houghton Colliery
Pontefract (Prince of Wales) Colliery.

(Collieries in italics were situated within the area defined as the West Yorkshire Coalfield.)

The collieries of the South Yorkshire Coalfield became part of the new No.1 (Worksop), No.2 (Doncaster), No.3 (Rotherham), No.4 (Carlton), No.5 (South Barnsley) and No.6 (North Barnsley) Areas.

Of the 958 collieries which the NCB had inherited, over half of them were found to be in immediate need of attention if productivity and output were not to suffer. To this end the NCB set about reorganizing the industry. In the newly formed North-Eastern Area there were thirty

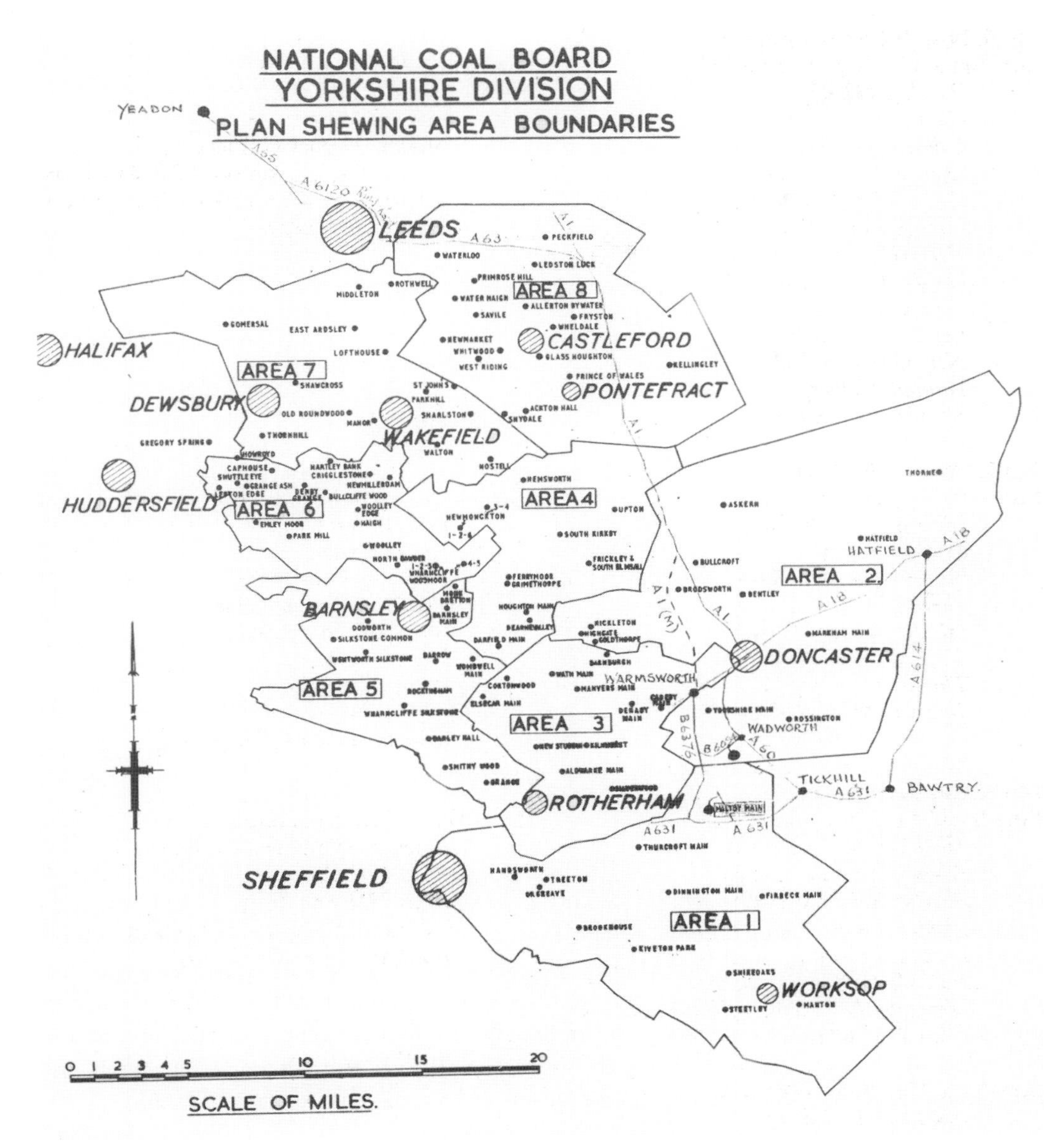

The eight Areas of the North Eastern (Yorkshire) Division, 1 January 1947.

major reconstructions on hand by January 1957, at an estimated cost of £65,161,000, including twenty collieries within the South Yorkshire Coalfield. These included the following schemes:

Colliery	**Estimated cost (£)**
Maltby	2,890,000
Brodsworth	4,044,000
Goldthorpe	1,089,000
Hickleton	3,992,000
Denaby & Cadeby Central Scheme	4,028,000
Manvers Central Scheme	7,441,000
New Stubbin	1,250,000
Silverwood	2,992,000
Darfield	620,000
New Monckton	6,980,000
South Kirkby	2,093,000
Grimethorpe/Houghton	6,931,000
Houghton (skip-winding)	301,000
Barrow	2,022,000
Dodworth	2,018,000
Wentworth Silkstone	499,000
Smithywood (virtually complete)	926,000
Woolley (sinking)	303,000
Woolley (coal preparation plant and Haigh surface conveyor)	738,000
Wharncliffe Woodmoor 4 & 5	1,137,000
Total estimated cost	**52,294,000**

'Story Of Reorganization Of Magnitude And Complexity', was the title of the second in a series of five articles called, 'Living on coal'. Published in the *South Yorkshire Times* on Saturday 6 June 1959, it featured the Rotherham Area. The twelve collieries making up the area had been operated by six different companies. Eight of the collieries were in production before the end of the nineteenth century. By the time of the article some twenty-four per cent of the area output was power loaded, Elsecar Main Colliery lead the way with eighty per cent. There was a Central Coal Preparation Plant and Central Workshop at Manvers and a second workshop at Elsecar.

The title of the third article was 'One Of The Smaller Areas In The North Eastern Division'. Published on Saturday 13 June 1959, it featured the North Barnsley Area. The area had been worked from the earliest times and many of the collieries were old and small. The larger collieries were situated in the South Yorkshire Coalfield proper, concentrated to the south around the town of Barnsley. The smallest colliery was Woolley Edge, with thirty employees, and the largest Woolley, with 2,350. The average seam thickness for the area was only 28in. The seams were of good quality, but produced a vast amount of methane gas. Woolley Colliery alone produced 3million cu.ft of methane per day. The area of coal between Woolley and Mapplewell, near Barnsley, was intensely mined. The report said: 'The area of coal worked in the Woolley/Mapplewell area is unique in Great Britain, here the coal is mined so intensely that no fewer than twelve seams are being worked at different levels at the same time by the three collieries, Haigh, Woolley and

North Gawber'. The complications arising in planning the underground workings, and dealing with the complex problems of strata movement, must have been formidable.

The last article in the series was called 'Probably The Best Reserve Of Coking Coal In Britain', Published on Saturday 8 August 1959, it discussed the importance of the mid-most area of the South Yorkshire Coalfield, the No.4 Carlton Area. The area to the north-east of Barnsley, worked by the New Monckton Collieries, contained some of the most valuable coking reserves in Europe. One effect of nationalization was to regard the working of British coal reserves in a holistic manner, aiming to work the reserves of coal in the most effective manner from the collieries within the area, rather than regarding the collieries as individual units working their own 'patch' of coal. Old, and out-of-date collieries were closed, development schemes were carried out at existing collieries and new collieries were sunk. In January 1947, the first month of nationalization, for the No.4 Area, the face output per man-shift was about 60cwt, and total output per man-shift about 20cwt. In February 1959 the figures were 78.7cwt and 28.2cwt, showing a vast improvement.

In the years after the Second World War, which coincided with the first decade of nationalization, energy was chronically short in Britain and Europe as a whole. The British coal industry of 1947 had a long history of contracting markets, falling output and bitter internal relationships. Starved of capital during the 1930s and during the Second World War, when continued private ownership was uncertain, the result was an industry which was badly under-capitalised. Expansion to meet the chronic energy shortage was difficult and slow. In the twenty years up to 1957 the needs of war and post-war recovery stretched the mines of Britain to the limit. The 'fuel gap' meant there was unending pressure on the industry to produce more coal and produce it at any price!

In the period from 1947 to 1956 the NCB spent £558million gross, £171million of which was spent on colliery schemes, principally major colliery reconstructions and new sinkings. In 1947 power-loaded output at the coalface was a mere two per cent of the total. By 1956 this had increased to more than fifteen per cent and by mid-1957 it was twenty-three per cent of total output. Productivity increased substantially and the British coal industry was producing cheaper coal than any other European producer.

Output in 1947 was 184m tons, and the NCB's first investment plan, the Plan for Coal (1950) aimed to raise output to 240m tons (later revised to 250m tons) by 1970. By 1957 the NCB had set about reorganizing and reshaping itself to meet further growth in demand. It was at this time that cheap and plentiful supplies of oil began to flow and the NCB was faced with the need to take drastic measures. For the first time in its history it began to close pits before their reserves were exhausted. What was to follow was a period of sustained contraction, lasting from 1957 until 1973. There was now too much coal and output targets were reduced. Between 1958 and 1959 eighty-five collieries, employing nearly 30,000 men, were closed.

It was the hidden potential of oil that had confounded the forecasters in the late 1950s. In 1957 a dramatic change in the market was to shatter the widely held belief that the rising trend in coal consumption would continue indefinitely. Between 1957 and 1959 the consumption and export of coal had contracted by 33m tons, and pithead stocks were at a peak of 36m tons. In the same period consumption of fuel oil increased to around the equivalent of 24m tons of coal. Competition from oil was here to stay. By 1964 sales of fuel oil had reached the equivalent of 55m tons of coal.

In 1956 the industry produced 207m tons, from 840 collieries with a manpower of about 700,000. By 1971 the output had fallen to 133m tons, the number of collieries had dropped to 292, and manpower was less than 290,000. Despite a partial recovery in the period 1962-1965, mainly due to big improvements in operational efficiency, the relentless long-term decline of the industry continued. By the mid-1960s the competition from oil intensified, and the government decided that it would not subsidise the industry. Coal had to compete, and to do so the NCB was forced to close pits faster than at any previous time. In the 1960s almost half of the British coal industry was abandoned as uneconomic. As a result the industry was virtually lost forever. In 1960 there were 698 collieries, by 1965 this had reduced to 534 collieries, and by 1970 the number had dropped to 299 collieries.

The Organization of Petroleum Exporting Countries (OPEC) was set up in 1960. Their aim was to restrict the flow of oil to maintain the price. They realized that the oil beneath their land was a finite resource and would have to be managed carefully to maximize its contribution.

In October 1960 Alfred Robens, later to become Lord Robens, joined the NCB as Chairman-designate. He quickly confirmed that he wanted to build the industry around an output of 200m tons per annum and the way to do this was to concentrate output in the best pits in the most productive coalfields. During Robens's ten year stint with the NCB productivity increased by seventy per cent, but at a cost to the industry. In 1959 some 700,000 men produced an output of 190m tons from 700 collieries. In 1969, after the greatest period of contraction of the industry in its history, the workforce had dropped to 300,000, producing 160m tons from a little over 300 collieries. Between 1957 and 1963 a total of 264 collieries had closed.

Despite the contraction, in the mid-1960s the coal industry was still a major industry, employing five to six times as many employees as ICI, the second largest British company and one of the largest companies in Western Europe.

A roll call of the economic, doubtful, and (probably) doomed pits

In November 1965 the NCB announced an accelerated pit closure programme affecting 120,000 men, one third of the labour force. Collieries were categorized as class 'A' (likely to continue), class 'B' (future doubtful), and class 'C' (likely to close). The plan called for the closure of 150 class 'C' collieries within the next two to three years. Of this number, ninety-five had almost exhausted their reserves, and the remaining fifty-five were said to be clearly uneconomic. For the most part class 'C' collieries were those described as 'gross losers' in the Government's White Paper on Fuel Policy. This left 281 class 'A' collieries with an assured future, and eighty class 'B' collieries categorized as doubtful, for geological, economic or labour shortage reasons.

In South Yorkshire seven collieries were classified as class 'B' and seven as class 'C'.

Class 'B' – future doubtful	**Class 'C' – likely to close**
Bentley	Barnsley Main
Denaby	Handsworth
Dinnington	New Monckton Nos 1 & 2
Hatfield	Smithywood
Kilnhurst	Wharncliffe Woodmoor 1,2 & 3
New Monckton	Wharncliffe Woodmoor 4 & 5
Thurcroft	Woolley Edge

Despite the contraction in the coal industry, the position of the Yorkshire Coal Industry appeared favourable. On 19 March 1965, the *Colliery Guardian* reported, in an article titled, 'Yorkshire to become Nation's Powerhouse', that there had been 'a revolution in the position of the coal industry over the last three years from a dead and dying one to a virile enterprise (Marketing Director for the Yorkshire Division).'

Natural gas from the North Sea, which was first proved in September 1965, made rapid progress. By 1967 it was contributing 36.4m tons equivalent of energy, replacing coal gas and providing a cleaner source of fuel for domestic and industrial use. By 1970 25m tons of coal were replaced by natural gas. North Sea oil, first proved in 1969, took longer to make an impact on the British energy scene. It was not until 1976 that sizeable quantities of North Sea oil came ashore. Nuclear energy was a little slower on the scene. The 17 October 1956 saw Queen Elizabeth II switch power from Calder Hall, at Windscale in Cumbria, into the National Grid, heralding the world's first commercial generation of nuclear energy. Fears for the long-term risks associated with its use and concerns about the true cost of this energy source, held back its large-scale use in Britain.

In the White Paper on Fuel Policy (1967) the government stressed that it was moving from a two-fuel economy, based on coal and imported oil, to a four-fuel one, with the addition of natural gas from the North Sea and nuclear energy. The White Paper also scaled back the Plan for Coal forecast of coal production for 1970 from the original 250m tons to 152m tons. In 1969, for the last time, coal outstripped all other energy resources combined, to provide 50.4 per cent of the nation's primary fuel intake. By 1971 oil had pushed coal into second place, and by 1974 coal's share of the UK market was down to one third.

NCB Reorganization 1967

In March 1967 the NCB's streamlined administrative structure was brought into operation and the Divisions and Groups ceased to exist as managerial units. The NCB was organized into three levels: National Headquarters, Areas and Collieries. The organizational changes were accompanied by a change to method-management by objectives. The resulting seventeen new-style Areas were each the size of a large business, employing over 20,000 men on average, with capital resources of £50million, and a turnover greater than £50million per annum. The reconstructed industry planned for 310 collieries by 1971, capable of producing as much coal as the coal industry of 1947, when there had been almost 1,000 collieries. The NCB's aim was to match the advances made in mechanization, with an infusion of best-practice management skills tailored to the structure and size of the industry.

Administrative reorganization of the Yorkshire coal mining industry was completed on 27 March 1967. From this date the existing seven areas of the Yorkshire Division ceased to exist, and four new self-contained areas were created – Barnsley, Doncaster, North Yorkshire and South Yorkshire. The four areas comprised ninety-two collieries with a total manpower of 94,000. The report on the Yorkshire Division, in the *Colliery Guardian* of January 1967, said the Yorkshire coalfield had the biggest potential of any in the UK, with proven reserves in excess of 4,000m tons.

The South Yorkshire Area was formed from the amalgamation of the former No.1 (Worksop) and No.3 (Rotherham) areas. The new area comprised twenty-two collieries employing over 34,000 men, with an output in excess of 9m tons per annum. Located within

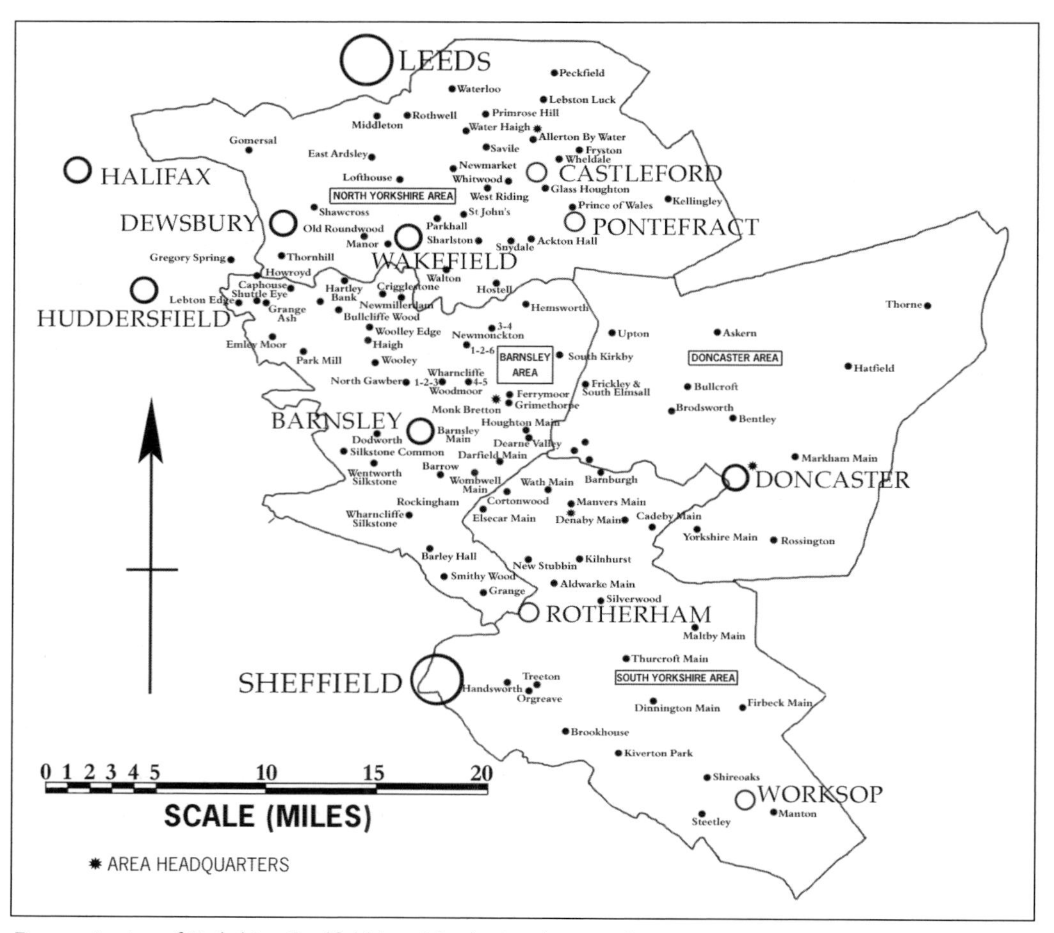

Reorganisation of Yorkshire Coalfield into North, South, Barnsley & Doncaster Areas, 27 March 1967.

the area were Brookhouse, Handsworth, Manton, Steetley, Dinnington, Kiveton Park, Orgreave, Thurcroft, Firbeck, Maltby, Shireoaks, Treeton, Barnburgh, Denaby, Manvers, Wath, Cadeby, Elsecar, New Stubbin, Cortonwood, Kilnhurst and Silverwood Collieries. The area also contained Manvers Coal Preparation Plant, the largest of its kind in Western Europe, and Manvers coking and by-product plant. The area headquarters were located at Manvers.

The Doncaster Area had the least number of producing collieries, thirteen, which employed 22,000 men and mined over 8m tons per annum. Located within the area were Rossington, Markham Main, Goldthorpe, Highgate, Hickleton, Askern, Hatfield, Yorkshire Main, Bentley, Bullcroft, Brodsworth, Frickley and South Elmsall Collieries. The area was acknowledged as having some of the richest coal-seams in the country.

The new Barnsley Area was formed from the existing Nos 5 and 6 (South and North Barnsley) areas of the North Eastern Division. The headquarters were located at Grimethorpe. Most of the collieries, in the former No.6 North Barnsley Area, were originally classified as being located in West Yorkshire. These include, Bullcliffe Wood, Caphouse, Crigglestone, Denby Grange, Emley

Moor, Haigh, Hartley Bank, Newmillerdam, Park Mill and Shuttle Eye Collieries. The collieries in this area were predominantly smaller in size than those further to the east. They totalled some thirty collieries, employing 25,900 men and producing 9.6m tons of coal per annum. By 1974 Barnsley Area output had dwindled to 6.1m tons per annum from a labour force of 16,330 men, and was still declining. Most of the twenty collieries remaining at this time were old. Many had been sunk between 1870-1890 and had exhausted their better seams. In 1974 plans were initiated to reconstruct the area, raising output to 8.5m tons from a labour force of 12,500. Between 1979 and 1984 the area was reconstructed and three major complexes created. The collieries in the area were connected underground and their output conveyed to the surface at the three complexes via inclined drifts. The three complexes were: the South Side, centred on Grimethorpe Colliery; the West Side, centred on Woolley Colliery; and the East Side, at South Kirkby Colliery. At each site a high capacity computerized coal preparation plant processed the coal, the one at Grimethorpe being one of the largest in Europe. The scheme cost £455million to implement. It was one of the largest reconstruction programmes in the history of British mining.

The North Yorkshire Area was formed by combining the former No.7 (Wakefield) and No.8 (Castleford) areas. The new area had twenty-eight collieries and employed over 22,000 men.

By the late 1970s the general economic climate had worsened, demand for coal from the CEGB was down and British Steel was taking less coke. The recession was more severe than expected.

In the 1950s miners' weekly earnings were twenty-five per cent higher than workers' in the manufacturing industry. From the late 1950s redundancies increased, fuelled by pit closures and increased mechanization (which increased from five per cent to over ninety per cent by 1970-1971). The psychological pressure on the industry to inhibit wage rises, together with the move into a day system of wage payments, resulted in a decline in earnings which started in 1958, and had become serious by the beginning of the 1970s, resulting in the coal strike of 1972. The strike lasted seven weeks and resulted in the miners winning more than two thirds of their pay demand. Twelve major power stations were shut down, 1.4 million workers were idle and there was talk of bringing in the troops to prevent a total industrial shut down. The famous 'Battle of Saltley' took place at the large Birmingham gas works, when pickets successfully closed the works. Arthur Scargill, the Yorkshire miner's leader, was projected to national fame during the strike.

The 1974 Plan For Coal

Between 1970 and 1974 the price of oil increased six times. The Arab-Israeli war of October 1973 and accompanying oil crisis, saw the first great rise in the price of oil. Within months the price of oil quadrupled. There was a growing realization of the danger of over-dependence on imported fuel and the value of home produced fuel resources. This was the background to the 1974 Plan for Coal, whose objective was to revitalize the industry by a massive investment to develop the nation's fuel resources.

The original concept of the plan was that between 1974 and 1985 the industry would replace some 40m tons of obsolete capacity with approximately the same quantity of new capacity, in order to maintain the output of the industry at the same level of some 120m tons. At the same time as capacity was replaced, it was planned to increase the output from opencast mining to 15m tons. The aim of the plan was to create a cost-effective industry. The report stated, 'an efficient competitive coal industry has an assured long term future.' The notion of the plan was to replace output from an 'ageing' industry (the average age of collieries was in excess of seventy years), but in an economic way. Between 1975 and March 1983 some £3,000million were to be invested in the industry

The adoption of the 1974 Plan for Coal was critical for the South Yorkshire Coalfield. It was the impetus behind the reorganization of the Barnsley area in the period 1979-1984. Problems surfaced during the reconstruction programme, British Steel was to have been one of the area's largest customers, but that market collapsed. One by one the collieries in the area closed.

To speed and smooth the flow of coal to power stations, rapid loading facilities were being introduced at collieries. In July 1971 the *Colliery Guardian* reported that rapid loading schemes, each costing £450,000, to fill 1,000ton liner-trains for the CEGB, were ready to start at Rossington, Brodsworth and Hatfield Collieries in the Doncaster area.

During the 1970s and 1980s coal was the main source of fuel used for the generation of electricity. Most of this coal came from UK mines, a resource-base which was plentiful, if not always the cheapest. In 1980 the CEGB, the NCB's largest customer, agreed to take 75m tons per year over the following five years, and British Steel agreed to limit coke imports to 4m tons

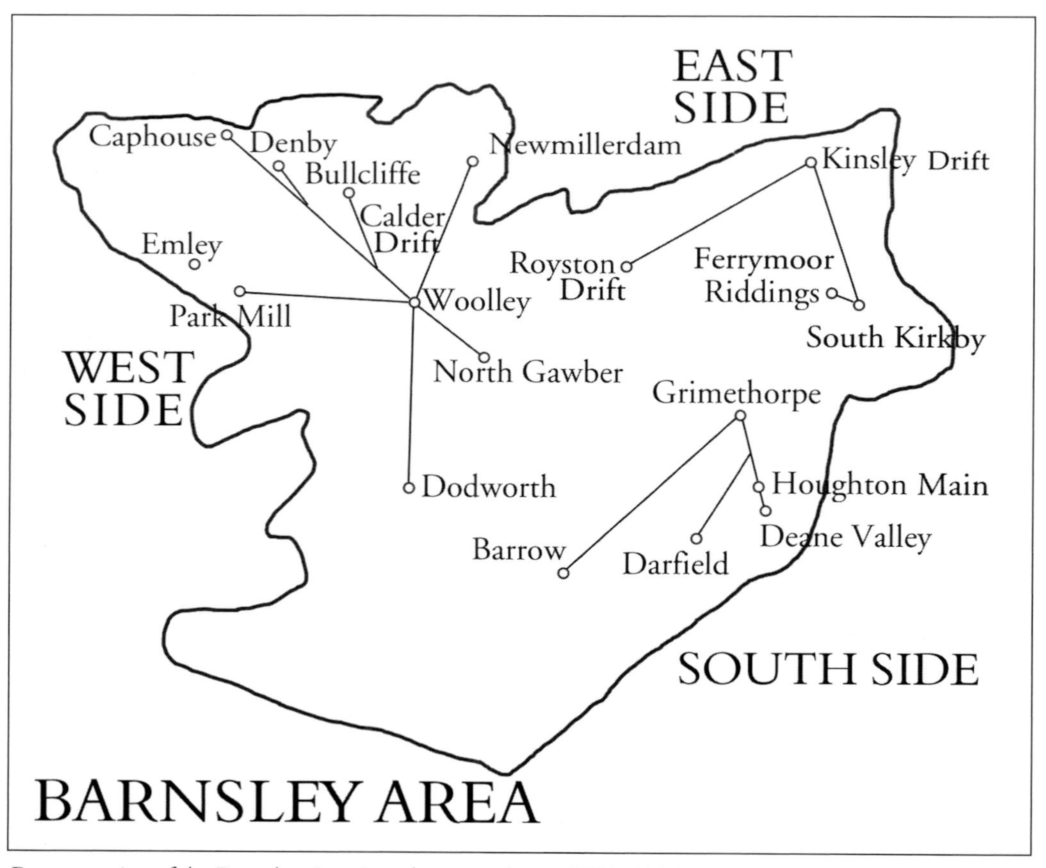

Reconstruction of the Barnsley Area into three complexes, 1979-1984.

that year. During the 1990s the pattern of fuel use change radically, with gas figuring for the first time in large-scale power generation.

The period from 1974 onwards can be described as a period of energy-diversification, as users were forced to examine alternative sources of energy supply.

Internal disturbances in Iran, followed by the revolution of 1979, further added to the West's unease about relying on imported fuel from unstable areas of the world. The electricity generation industry was the NCB's largest customer, taking three quarters of output. However, demand for coal for electricity generation was down and alternatives, such as nuclear energy, were on the increase. Only one coal-fired power station, Drax 'B', had been ordered since the 1960s.

With a substantial price advantage over oil, the expansion in the use of coal for industry was expected to treble in size, and perhaps increase by up to five times. For a number of years the NCB had been developing what was known as 'fluidized bed combustion'. Potentially there were enormous benefits to this type of boiler; the system cost was lower, very poor grades of fuel could be burned and the system could be adapted to reduce sulphur dioxide pollution. The benefits, if they could be realized, were enormous. In 1980 a large test plant was being commissioned at the Grimethorpe site to assess the potential of fluidized bed combustion for electricity generation.

The period 1979-1981 saw the emergence of a world energy glut, which resulted in a collapse of world coal prices from around $70 to $40 or less per metric ton. Around the same time that the British coal industry reached a surplus of output, so did the rest of the world! This put tremendous pressure on coal prices. The NCB was threatened with the erosion of its market to its two major customers, the British Steel Corporation and the CEGB.

The publication in 1980 of a massive report on the future of coal, argued that coal production world-wide would increase by up to 200 per cent by the end of the century. Coal would be a 'bridge' to the future. British coal costs were the cheapest in Europe and the industry received the lowest subsidies. In 1979 the cost of production of a metric ton of coal was £58 in Belgium, £45 in France and £41 in West Germany, while in Britain it was £29. The government refused to ease the cash limits on the industry. Their aim was to make the NCB break even by 1984. Jobs and pits would inevitably have to go. By February 1981 Derek Ezra, the NCB Chairman, told a meeting of the NUM that twenty to thirty pits were to go, and that this number might ultimately be as high as fifty.

In the mid-1980s the coal industry faced problems related to the halving of oil prices, the fall in international coal prices, and the strength of the pound sterling against the US dollar, which made foreign coal and oil cheaper for British customers to buy. In 1983-1984, the year before the national mining strike, the industry lost £875million.

The 1984-1985 Coal strike

Much has been written about the 1984-1985 coal strike, so I will not dwell on this subject other than to point out that the strike was a reaction to the announcement of the proposed pit closure programme. One of the central arguments was the existence of a NCB 'hit list' of collieries marked for closure. The militants in the industry argued that the list was a death warrant for the industry. The NCB insisted that no such list existed and that closures were negotiable. Within the industry there was the much-vaunted figure of 'twenty pits and 20,000 jobs'. When the NCB Chairman,

Ian MacGregor, presented his 1984-1985 budget for the industry on 6 March, he let it be known that he was looking for an output reduction of some 4Mtons, the same as the previous year, when 20,000 jobs had been lost! MacGregor's style for trimming the industry was to present his twelve area directors with strict budget limits and leave them to make the cuts.

In South Yorkshire the crux of the hit list, real or imaginary, was the closure of Cortonwood Colliery. The area director for South Yorkshire believed he could make the required savings, amounting to 500,000 tons, by the closure of just one colliery. The reasoning was that Cortonwood, sunk in 1873, was producing a grade of coking coal for which there was no demand. Work could be found for those displaced at nearby pits. What the area director had not realized, was that this was just what the militant NUM had been searching for. Cortonwood was not a dying colliery, it had a good production record and its losses were small. In addition, the men had been told that the colliery's future was secure for the next five years. Now they were being told the pit would close in five weeks. If the Board could close Cortonwood at such short notice, NUM leaders argued, they could close any pit – none were safe. This was the decisive propaganda that the NUM wanted. The strike started on 12 March 1984 and lasted twelve months. It was the culmination of the longest and bitterest industrial dispute in British history. It cost one murder, thirteen further deaths, thousands of injuries, almost 10,000 arrests and more than £7billion of taxpayer's money. After the end of the strike,

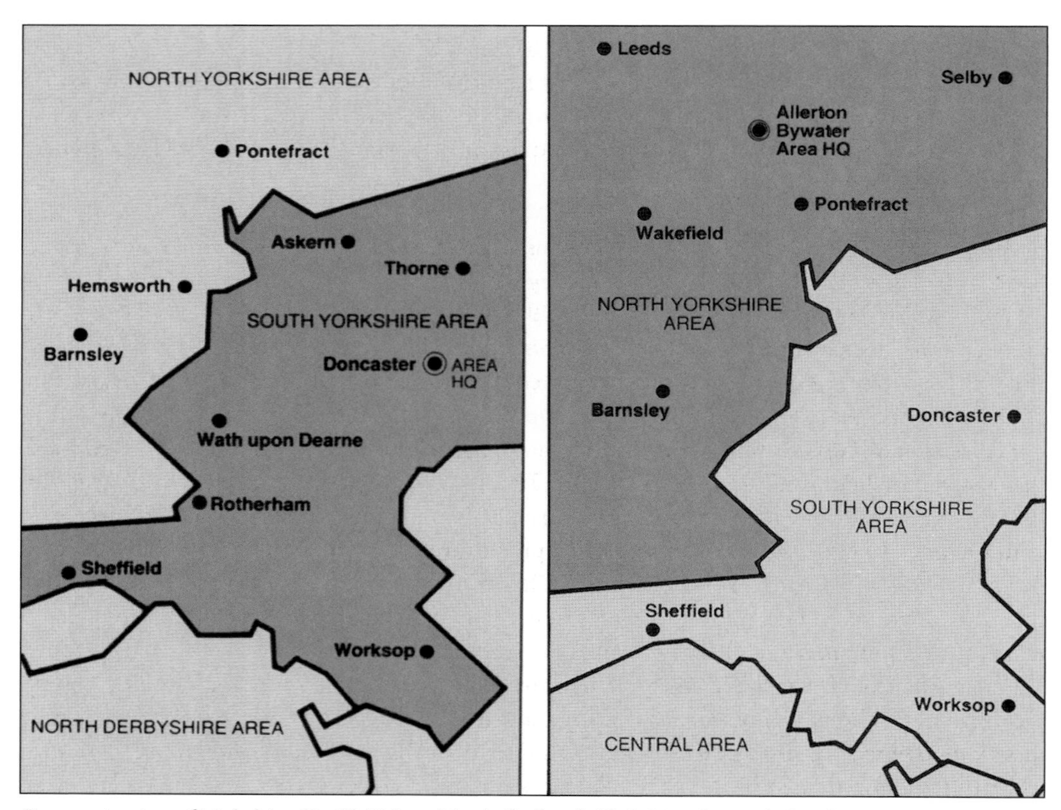

Reorganization of Yorkshire Coalfield into North & South Yorkshire Areas, 1 October 1985.

in March 1985, a large number of collieries closed. The implications for Britain's economic, political and social structure over the coming decades would be far-reaching.

Following the national coal strike an intensive effort was made to win back customers and to find new business for the industry. On 27 March 1985 a massive reorganization of the Yorkshire Coalfield was announced, the first since 1967. It came into effect on 1 October 1985, when a new South Yorkshire Area was formed, by the amalgamation of the old South Yorkshire and Doncaster Areas. The Barnsley and 'old' North Yorkshire Areas merged to form a new North Yorkshire Area. On 2 October 1985 the *Daily Telegraph* reported:

> ***8,000 Yorks pit jobs to be lost voluntarily.***
> *...The reorganization is expected to make large financial savings for the Coal Board, which sees the programme in line with its campaign for greater efficiency...*
> ***Output targets***
> *Three pits in South Yorkshire are scheduled to close, and the management in both areas want other pits to improve their output performances to meet the Board's new productivity target of £39 a tonne. Emphasis on production was moving from collieries in the west of Yorkshire to those in the east. The new Selby Complex would be four times more productive than the collieries of today.*

On 21 December 1985 the *Daily Telegraph* reported:

> ***23 COLLIERIES HAVE CLOSED SINCE STRIKE***
> *Twenty-three collieries have closed and 18,500 miners have taken redundancy since the coal strike ended in March, Mr Hunt, Energy Under-Secretary, said yesterday.*
> *He added that seven National Coal Board workshops, a tar plant and coke works had also closed since the end of the strike.*

In the 1980s the South Yorkshire Area was consistently producing the cheapest coal in the country, and was described as 'the coal industry's jewel in the crown'.

On 5 March 1987 the Coal Industry Act 1987 received Royal Assent. The NCB was no more and the British Coal Corporation was formed to take ownership of the coal industry. In 1987 almost 10m tons of coal were imported. This included 3.4million from Australia, 2.9million from the USA, 1.1million from Poland and 0.7million from the Netherlands.

The electricity industry was privatized in the late 1980s. This resulted in an increase in imported coal and reduced demand for indigenous coal. The result was that pit closures started again. In January 1992 there were just fifty-three working collieries remaining in the coalfields of Britain.

In October 1992 British Coal announced a pit closure programme in which thirty-one pits had been earmarked for closure, with ten collieries for immediate closure. Included in the shortlist of ten were Grimethorpe, Houghton Main and Markham Main Collieries, all of which were profitable at the time.

It was anticipated that the industry could only secure a contract for 40m tons of coal from the English and Welsh electricity power generators for the year starting April 1993, and that this would fall to a maximum of 30m tons in subsequent years. This problem of over-capacity was further compounded by the high, and rising, level of stocks, which stood at 46m tons at pitheads and power stations.

Following the announcement of colliery closures, marches and rallies were held in support of the miners. Two major marches were held in London on 21 and 25 of October. The situation did not improve and in December 1993 the *Coal News* reported:

> ***THE FACTS BEHIND THE PIT CLOSURES***
> *Against a background of increasingly serious imbalance between coal demand and supply, British Coal has proposed the closure of collieries in the North East, Yorkshire, Nottinghamshire and Staffordshire Coalfields.*
> *There will be manpower and output reductions at other pits as the prospects for increased sales to the electricity generators this year diminish against this background that coalfield mining union representatives have been told there is no market for the output for eight pits – irrespective of the cost of production.*

Hatfield, Bentley and Frickley were three South Yorkshire collieries on this list of eight collieries to close.

On 5 July 1994 the Coal Industry Act 1994 received Royal Assent for all but a tiny part of the British coal industry, signalling the end of almost forty-seven years of state control. The Act provided for the setting-up of the Coal Authority in Mansfield, Nottinghamshire, who took ownership of the coal reserves in and off Great Britain with effect from 31 October 1994. Nineteen collieries and thirty-two opencast sites were transferred into private ownership. RJB Mining took seventeen collieries and fourteen opencast sites. Celtic Energy took control of nine opencast sites in South Wales and Mining (Scotland) took control of Longannet Colliery, the last deep mine in Scotland, and nine opencast sites. In South Wales an employee buy-out took control of Tower Colliery, and Coal Investments took leases and licenses on two pits in Yorkshire and two in the Midlands.

On 30 December 1994 Britain's core mining assets, comprising deep mines and opencast sites, returned to the private sector. The new owners paid almost £1,000million to the Government for the most modern mining business in Europe. Nineteen deep mines, thirty-two opencast sites, twenty-eight disposal points and over 700m tons of reserves were sold, together with contracts to supply 100m tons of coal to the electricity generators in England, Wales and Scotland, by March 1998. Prior to the sale, ten collieries not required by British Coal had been acquired by the private sector under lease and license. The future of two other collieries remained to be resolved. There were a further 120 small mines, never owned by British Coal, which produced about 1m tons a year. Following the sale of the core mining assets, the Coal Authority, an arm of the Government responsible for the management of the nation's reserves of unworked coal and the licenses and leases required to work it, was set up.

Of the surviving collieries in the South Yorkshire Coalfield, RJB Mining purchased Maltby and acquired Rossington under lease and license. In mid-1996, after running the colliery for two years, RJB Mining purchased Rossington. Markham Main Colliery was purchased by Coal Investments PLC and Hatfield Colliery was purchased by a management-led buy-out, and run as the Hatfield Coal Co.

In 1997 the UK market for coal stood at 58m tons. By 1998 it had dropped to 29m tons and looked set to decline further.

Development Phases Of The South Yorkshire Coalfield

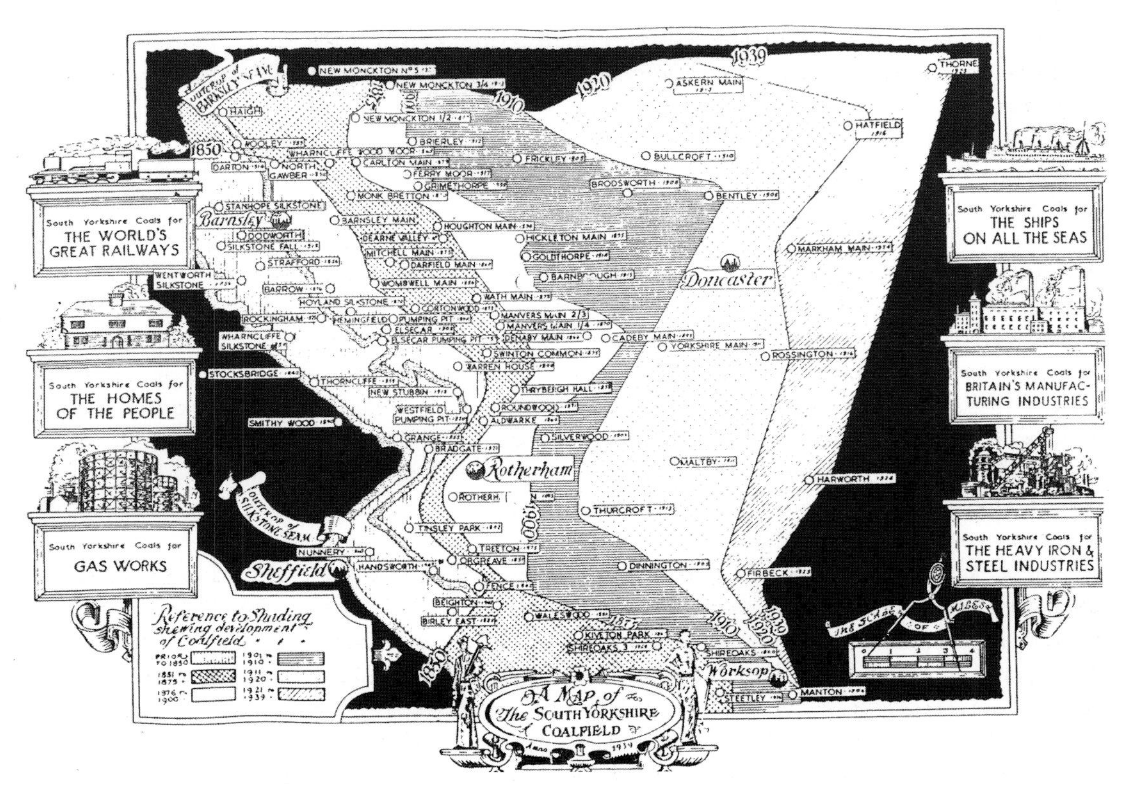

The above map shows the development of the coalfield, from the outcrop of the major seams on the foothills of the Pennines, in a relentless easterly direction towards Doncaster and beyond. The map clearly illustrates the density of mining in the central portion of the coalfield.

The progress of mining, in an eastwardly direction, by the key dates; 1850, 1875, 1900, 1910, 1920 and 1939 is shown by the contour lines. For the purpose of the profiles, individual collieries have been grouped by the geographical area of the phase into which they are located, irrespective of whether they were sunk at the time of the phase or at a later date. This is particularly applicable to collieries sunk in the geographical area covered by the phase of mining before 1850, where sinkings, to work the deeper seams, were taking place well into the early 1900s

6
Mining Prior To 1850

It is likely that the Romans worked coal along the outcrop of the coal-seams, but the first recorded attempts to work the seams of South Yorkshire came in the medieval period. In the thirteenth and fourteenth centuries there are various records connected with mining, usually recording the death of some unfortunate miner. A number of leases and grants for coal working near Wentworth were granted to the Fitzwilliam family, *c*.1370.

By the thirteenth century most of the British coalfields were being worked, and South Yorkshire was no exception. Coal was mined for local use, producing a few thousand tons at most. Mining at this period was frequently a seasonal occupation, occupying the time when men were free after harvest, and before the roads and tracks had deteriorated too much with the winter weather.

It is well-known that during the period 1550-1650 there was a minor 'industrial revolution' in the coal industry of Britain. Output for the country as a whole is estimated to have increased from about 200,000 to some 3m tons.

Between 1550 and 1700 there had been a great expansion in the use and mining of coal and ironstone to feed local industries. This early phase of development was confined to a band along the western side of the coalfield, following the outcrop of the Barnsley and Silkstone Seams and accompanying ironstone bands, particularly the Tankersley and Silkstone ironstones. By the mid-eighteenth century a thriving ironstone industry was operating between the outcrop of the coal-seams, running diagonally from Bretton, near Barnsley in the north, to Masborough, near Rotherham, in the south. A string of furnaces, forges and slitting mills operated on, or near, this band of ironstone mining activity. In addition there were references to mines and coal workings at Cudworth, Ardsley, Thurnscoe, Rotherham and Sheffield, in areas away from the outcrop.

The early mining of both coal and ironstone was carried out at, or near, the outcrop, where the coal and ironstone was relatively accessible. Mining by drifts, bell-pits and shallow shaft workings would have gone on at this period. The remains of many bell-pits can be found at a number of locations, such as on either side of the M1 motorway near Hoyland village, south of Barnsley.

Coal mining in seventeenth century South Yorkshire was a 'land-sale' industry. At this period there were no navigable rivers or canals, and no sea-ports. Coal would have been consumed locally, over a relatively small area.

During the eighteenth and nineteenth centuries the Industrial Revolution accelerated. Development of the coal industry was driven by the demand for coal (for industrial and domestic consumption), improvements in mining technology the channels of communication, which extended the market for South Yorkshire coal.

Developments in mining technology in South Yorkshire kept pace with other regions in Britain. The Newcomen engine, which enabled deeper seams to be worked, was introduced at a colliery near Chapeltown in 1753. South Yorkshire had its own pioneering mining engineer, in the guise of John Curr, Viewer to the Duke of Norfolk's Sheffield collieries. In 1787 Curr introduced wooden corves in the coal-drawing shaft, to replace the wicker baskets used to raise coal, and in 1788 patented shaft guides to simplify and speed coal winding. Curr also applied the steam engine to underground haulage in 1805, and introduced cages for use in shafts. Equally important was Benjamin Biram who, like his father before him, was Superintendent of Collieries to Earl Fitzwilliam of Wentworth. In 1824 Biram patented an anemometer, to measure the speed of the air-current underground, the design of which is little different from the design in use today. Biram patented a centrifugal ventilating fan in 1842, a number of which operated at the Earl's collieries at Elsecar. Biram's was one of the first centrifugal fans for which there is documentary evidence, and was certainly effective at a time when the South Yorkshire Coalfield was suffering from the development of the gassy Barnsley Bed Seam. Biram fans were still at work in the early years of the twentieth century. In 1852 Biram placed one of his fans underground at Hemingfield Colliery, near Elsecar, the first occasion that a ventilating fan had been sited underground.

Between 1750 and 1850 the metal smelting industry in the area expanded, creating a great demand for coke. Companies like Samuel Walker of Masborough, near Rotherham, and Newton Chambers of Chapletown, near Sheffield, developed rapidly in this period.

During the Turnpike era of 1740-1826, better roads were built in South Yorkshire, and by the mid-eighteenth century the Don Navigation had been improved as far as the outskirts of Sheffield. The Dearne & Dove Canal Bill was passed in 1792. The canal, and its branches to Elsecar and the Dove valley, had been constructed by 1798 and were operational by the early 1800s. By 1804 canals linking the rivers Calder, Don and Trent had been constructed. The development of the road and canal infrastructure saw the opening-up of more collieries, as the transport system provided a ready outlet route for the coal. The development of the railway system, from the mid-nineteenth century onwards, further improved transport and provided a convenient means of rapidly moving vast and bulky cargoes of coal. In 1835 Sheffield consumed 515,000tons of coal.

By 1850 Barnsley could be said to be the 'coal capital' of South Yorkshire, with a cluster of collieries surrounding and within the town. The majority of the collieries at this time were situated between the outcrop of the Barnsley and Silkstone Seams, the two seams providing half of the total output of the region. At this date a small number of collieries had been sunk a few miles to the east of Barnsley. The deepest collieries of the day were about 200yds deep.

As well as the coal mining development along the outcrop of the Silkstone and Barnsley Seams, it should be noted that there were a large number of small-scale enterprises working along the outcrop of the Gannister Coal, to the west of the Silkstone Seam outcrop. A thriving small-scale industry, mining thin coal-seams, clay and gannister (a type of fireclay) developed.

Before 1850 both the Frickley and Maltby Troughs, where the coal was deeper, were avoided for a combination of technical and financial reasons, and no collieries had been sunk in the concealed region of the coalfield.

In the mid-nineteenth century the West Yorkshire Coalfield was more important than the South Yorkshire Coalfield. West Yorkshire produced some 5m tons per annum from about 250 collieries, compared to South Yorkshire's output of 2.8m tons.

The majority of the collieries sunk within the area covered by this phase of mining, and which survived until nationalization, were sunk after 1850 to work seams other than the Barnsley Seam, which was worked out over much of the area by this time. An exception was Barnsley Main, whose ancestry can be traced back to the Oaks sinkings. Some of these collieries were located at, or close to, the site of earlier collieries. Many of the collieries in this phase were linked to, and used, earlier colliery sites for ventilation and pumping.

Barley Hall Colliery, April 1979.

Collieries located in this phase of development include:

- Barley Hall
- Barnsley Main (including New and Old Oaks Collieries)
- Barrow
- Brookhouse Beighton
- Darton
- Dodworth
- Elsecar Main
- Grange (including Bradgate Drift)
- Haigh (drift mine complex)
- Handsworth Nunnery
- Hoyland Silkstone
- New Stubbin
- North Gawber
- Nunnery
- Rockingham
- Stocksbridge
- Thorncliffe
- Thorpe Hesley
- Wentworth Silkstone
- Wharncliffe Silkstone

Barley Hall Colliery

Situated about half a mile to the west of Thorpe Hesley village, the colliery took its name from a property of the same name, located 200yds to the north-east of the colliery.

Sinking of a single 12ft-diameter shaft, on land leased from Earl Fitzwilliam, commenced on 21 June 1886, reaching the Silkstone Seam at 165yds on 25 January 1887. The rate of progress of the shaft-sinking was believed to be a record at the time. When sunk the pit was used as the upcast ventilation shaft for the Norfolk & Smithywood Collieries, located to the south-west of Barley Hall. In 1914 Newton, Chambers sank Thorpe Pit on the edge of Thorpe Hesley village, whose two shafts served as ventilating, pumping, man-riding and service shafts for Smithywood colliery. The Barley Hall shaft then reverted to Thorncliffe colliery, serving as the upcast and service pit and for pumping water from the Norfolk workings.

One section of the Parkgate Seam workings was reached by a steep drift from the pit yard at Barley Hall. It passed under Hesley Lane, which ran along the side of the colliery. This was called the Wembley Drift because it was opened in 1924 at the same time as Wembley Stadium. Contrary to normal practice, the drift was driven upwards from the workings to the surface. A workman was sent to watch for the drifter's drill piercing the ground, as the drift approached the surface.

Coal was never wound at Barley Hall, originally it was brought out at Thorncliffe Colliery incline, and later via connecting roads, at Smithywood Colliery. During its life the colliery produced between 200,000 and 250,000tons per annum.

In February 1964 the *Colliery Guardian* reported, 'Rundown fears groundless'. There were fears at the time that Barley Hall would be run-down by 1966. The extra manpower, which the colliery had received as a result of the closure of Grange Colliery in the early 1960s, had substantially reduced the life of the remaining reserves. However, the boundaries of the colliery were realigned and the colliery was working the Fenton Seam, which had previously been the preserve of Rockingham and Elsecar Collieries.

In the year ending March 1973 the colliery had a manpower of 374 and produced an output of 232,000tons. The last shift was worked at the eighty-eight-year-old colliery on 31 May 1975, when it closed due to exhaustion of reserves.

After closure a scheme to develop a mining museum on the site was proposed. The *Colliery Guardian* reported in October 1975, 'Colliery may become Museum'. The colliery, complete with pithead gear, winding-engines, pump-house and surface buildings was offered to the South Yorkshire County Council. Though approved in principle, the scheme did not materialize and the surface was demolished.

Seams worked

Seam	Depth (yds)	Abandonment date
Top Fenton	52	May 1975
Low Fenton	54	May 1975
Parkgate	74	May 1975
Thorncliffe	92	January 1970
Top Silkstone	165	February 1968
Whinmoor	225	January 1958

Barnsley Main Colliery

Barnsley Main Colliery was situated in the borough of Barnsley, two and a half miles south-east of the town centre. The colliery had two shafts 400yds apart, the No.2 shaft was sited at what was known as the Hoyle Mill Side and the No.4 shaft at the Stairfoot Side. Originally each side of the colliery had four shafts, at the Hoyle Mill Side were the Nos 1, 2, 3 and cupola shafts, and at the Stairfoot Side, the No.4, Meltonfield Water Pit and the two Old Oaks shafts. The colliery was served by the LNER, Mexborough to Barnsley branch and the LMS, Cudworth to Barnsley branch.

The early history of the colliery is complex, not least by changes in name and ownership during its life span, particularly in the early days. Two shafts, eventually known as the Old Oaks shafts, were sunk in 1824 to work the Meltonfield Seam at a depth of 112yds. The two shafts were 7ft 6in in diameter and 25yds apart and both were downcast. One of them was used for coal raising and the other for pumping. In 1851 the shafts were deepened to the Barnsley Seam at a depth of 285yds.

The colliery operating on the Hoyle Mill Side utilized the No.1 shaft (known as the South Pit) as the upcast and coal winding shaft, and the Cupola shaft (known as the North Pit) as the downcast. The colliery, typical for the period, was ventilated by an underground furnace. The Barnsley Bed Seam, varying from 8-9ft in thickness, was worked. The No.4 shaft, at the Stairfoot side, was employed as a pumping pit. Coal was transported using the nearby Dearne & Dove Canal. From 1835 the colliery was owned and run by Firth, Barber & Co.

In 1838 two more shafts were sunk to the Barnsley Seam. These were on the Oaks property and were known as the Ardsley Main shafts. One of them became the upcast for the Old Oaks Colliery.

On 5 March 1847 an explosion in the Barnsley Seam resulted in the loss of seventy-three lives. Following the explosion the colliery was virtually closed, and during this time a large

Barnsley Main Colliery No.4 shaft (Stairfoot Side).

stone fell down the No.1 shaft, causing serious damage. Soon afterwards the shaft was filled. The North Pit was retained and became an upcast for a new underground furnace. Two new shafts were sunk as downcast and coal winding shafts. Coal production recommenced about 1851 and was transported by rail.

In about 1858 a small staple-pit was sunk, about 1,000yds north-east of the Oaks Shafts, to prove a fault. During sinking a feeder of methane was struck at a depth of 12yds. A small gas-holder was erected over the pit, and the gas collected for use at the colliery.

On 12 December 1866 there occurred the first of thirteen explosions, which took place over ten days, resulting in the loss of 364 lives. The disaster was the worst in Britain at that time, and was not to be surpassed for many years to come. Following the disaster, the two shafts at the Oaks, which came to be known as the Old Oaks, were filled and new shafts were sunk at Ardsley, about three quarters of a mile from the old colliery.

Sinking of the new colliery, close to Stairfoot railway station on the Manchester, Sheffield and Lincolnshire Railway, commenced in July 1867, reaching the 8ft thick Barnsley Seam at 336yds in July 1870. Two 12ft-diameter shafts, 15yds apart, were sunk. Sinking was made difficult by the vast amount of water given off by the Oaks Rock, and the upper 150yds of each shaft were tubbed with cast-iron tubbing. During sinking a valuable seam of clay was discovered, which provided the raw material for brick-making at the colliery. The Oaks 'Hards', an excellent quality steam-coal, was in high demand by the London market, by railway companies, and for consumption in Sheffield. A large quantity was also exported, principally to northern Europe and India. The new colliery became known as the New Oaks, and was one of the largest in the West Riding of Yorkshire. The royalty for the combined Oaks Collieries amounted to almost 1,500 acres, and was served by the Hull & Barnsley and Manchester, Sheffield & Lincolnshire Railways. Seventy-two beehive type coke ovens were erected at the colliery.

In the early 1870s the No.4 pumping shaft was abandoned, and a new shaft, the Meltonfield Water Pit, completed in 1875, was used for pumping.

In 1889 No.4 shaft-end (what was later to become known as the Stairfoot End) was taken over by the Barnsley Main Colliery Co. The two sides were now separate collieries. From 1893 to 1896 the No.4 shaft was reopened, and sunk, with a 16ft-diameter, to the Silkstone Seam, at a depth of 647yds.

In about 1890 the No.2 shaft-end took the name Rylands Main Colliery (after Dan Ryland the owner of the nearby Rylands Glass Works). In 1890-1891 the No.2 shaft was deepened to work the Winter Seam. The colliery worked the Melton Field and Winter Seams, the Winter Seam was extensively worked until 1898.

In the late 1890s the No.3 shaft, which had the largest diameter, at 18ft, was sunk. On reaching the Two Foot Seam, a fall of stone damaged the shaft, and it was abandoned. At this time there were sixty-two beehive coke ovens in operation at the Old Oaks site. Also, following the death of Dan Rylands, the Barnsley Main Colliery Co. acquired ownership of the whole colliery. Though both sides became one colliery unit and were interconnected, each worked separate seams, and were complete with separate screening and washing plant.

The No.2 shaft was deepened to the Lidgett Seam, and further deepened to the Fenton Seam in 1916. The Fenton Seam was initially hand worked but, because of the strong nature of the seam, it was decided to install coal-cutting machines to undercut the coal, and reduce the workload. In 1924 both bar and chain type machines were tried, the chain type machine proving the most successful.

Barnsley Main Colliery No.2 shaft (Hoyle Mill Side), March 1988.

In 1929 the No.2 shaft-end (Hoyle Mill Side) was closed down due to the depression in the coal trade. Following the amalgamation of the Barrow Barnsley Main Collieries, in 1932, the No.2 shaft-end was re-opened in 1933.

In February 1942 explosions took place in the Fenton Seam, resulting in twenty-nine injuries and the loss of thirteen lives. Further problems, including an underground fire and difficulties encountered in working the Haigh Moor and Swallow Wood Seams, made the position at the colliery precarious.

The No.4 shaft-end (Stairfoot Side) continued to work the deeper seams until 1946, although serious damage had been caused to the shaft as a result of working the shaft pillars in the Fenton Seam.

The No.2 downcast shaft was 15ft in diameter and 512yds deep. From this shaft the Swallow Wood, Haigh Moor and Lidgett Seams, at depths of 344, 358 and 393yds respectively, were raised. The No.4 upcast shaft was 16ft in diameter and sunk to a depth of 640yds. From this shaft, the Fenton and Parkgate Seams, at depths of 524 and 542yds respectively, were raised. The colliery was now known as Barnsley Main Nos 2 and 4 pits, the No.2 shaft the upcast and No.4 shaft the downcast. In 1942 the colliery had a weekly target of 10,500tons.

From 1932 until nationalization the colliery was owned by Barrow Barnsley Main Collieries Ltd. In the process of reorganization the company decided to work Barnsley Main coal from their two adjoining collieries, Barrow and Monk Bretton. During 1945 the decision to close Barnsley Main was made, but on account of representations made to the company, it was decided to cease working in all the seams currently being worked and develop the Kent's Thick Seam at a depth of 213.5yds. On 1 December 1945 development work started from an inset in the No.2 downcast shaft. It was also the intention to work the Beamshaw and Winter Seams, at depths of

161.5yds and 150yds, by either short drifts or staple pits. Ultimately, it was intended to drive a drift from the Kent's Thick Seam to the surface, to convey coal direct to the washery. By 1 December 1946 1,000tons per day were being wound in a single shift.

Following nationalization a reconstruction scheme was carried out at the colliery. The scheme involved the reconstruction of the No.4 shaft pit-bottom, and erection of a new Baum coal washer, and associated railway sidings, at the surface. The improved coal preparation enabled poorer grades of coal to be worked. At Vesting Day in 1947 only one seam of coal was being worked, but by 1957 there were three seams in production. The reconstruction cost a total of £310,000.

In 1956 Nos 2 and 4 shafts were filled to the Haigh Moor Seam level, due to gas problems at the deeper seams. A surface drift was driven from the Beamshaw Seam at about this time to convey coal direct to the washery. The colliery continued to use the drift for conveying coal to the surface, men and materials via the No.4 shaft to the Beamshaw Seam, and other materials via the No.2 shaft.

In May 1966 Barnsley Main wound its last coal and the colliery closed as an independent producing unit. Men at the colliery were transferred to the surrounding collieries of Grimethorpe, Ferrymoor, Barrow and Darfield.

By the late 1960s Barrow Colliery was running out of coal and started to work coal in the old Barnsley Main Colliery take from 1971-1972. By June 1975 the last face in the original Barrow take had finished working, and from that time onwards all of the coal mined by Barrow Colliery was taken from the former Barnsley Main take. Entry into the old Barnsley Main workings had been beset with problems due to the proximity of old waterlogged goaves and old workings, both above and below the areas being worked. From the early 1970s the men working in the Upper Level (i.e. the Swallow Wood Seam) were transported overland by bus from Barrow Colliery to Barnsley Main, in order to eliminate the long underground journey. The shafts and surface at Barnsley Main were fully refurbished for man-riding and materials access, while coal winding and processing remained at Barrow Colliery.

Barnsley Main had been reborn under a £25million rehabilitation scheme, comprising new surface buildings and a refurbished shaft.

The Barnsley Main Colliery take had been extensively worked, coal having been extracted from all of the fourteen workable seams, seven of which were exhausted. The deeper seams, however, were more intact, having been less exploited, and it was to these that Barrow looked for its future production. By the mid-1970s a major development programme was required to give access to sufficient reserves of coal to ensure adequate bulk tonnage. The nearest working faces at the time were some three miles from the shafts and extended to six miles at the furthest point of the take. At this time the Lower Horizon (i.e. the Silkstone Seam) workings were extending in the general direction of Grimethorpe Colliery, and the possibility of making an underground connection with the Grimethorpe shafts, in order to wind and process the coal at Grimethorpe, was considered.

In the period 1979-1984 Grimethorpe Colliery became the hub of the £174million 'South Side' project complex of the Barnsley Area. Barnsley Main, Dearne Valley, Houghton Main and Grimethorpe were linked underground, and the output from these collieries surfaced at Grimethorpe, up a new 2.5km tunnel, for processing in the new central coal preparation plant.

By the mid-1980s seventy per cent of the colliery's output went to power stations. However, by the late 1980s Barnsley Main faced the problem of rapidly diminishing reserves, and the solution adopted was to employ short-life retreat face working. At this time the colliery was working the Fenton and Parkgate Seams, with a labour force of 500 men and had a production target of 700,000 tons in 1990. On 1 March 1991 the *Daily Telegraph* reported:

> ***700 to be laid off as pits close.***
> *More than 700 jobs are to be scrapped with the closure of two pits, British Coal announced yesterday... Barnsley Main lost £5 million this financial year and with rapidly worsening geology there was no prospect of any improvement, according to Group Director, Mr Bob Siddall. Some 470 miners are affected by the planned shutdown...*

As part of Barrow Colliery, workings in the Barnsley Main take finally ceased coal production on 19 July 1991.

Seams Worked:

Seam	**Depth (yds)**	**Thickness (inches)**	**Abandoned**
Beamshaw Seam	174	26	1967
Kent's Thick Seam	213.5	46	1967
Swallow Wood Seam	344	36-44	1944
Haigh Moor Seam	358	41-54	1944
Lidgett Seam	393	31-48.5	1944
Fenton Seam	524	52	1991
Parkgate Seam	542	42-48	1992

Barrow Colliery

Barrow colliery was situated in the township of Worsborough, three miles south of Barnsley. The royalties under lease to the company amounted to 3,500 acres, and had an average gradient of 1 in 12 in a north-easterly direction.

Towards the close of the nineteenth century large steel producing companies began operating collieries in the expanding South Yorkshire Coalfield, to provide them with coal and coke. The Barrow Haematite Iron & Steel Co. who produced iron and steel in the Furness region of north Lancashire (now part of Cumbria), was one such company. The attention of the company had been directed towards South Yorkshire because of the excellent coking qualities of the coal. Towards the end of 1872 the company purchased Worsborough Park Colliery, near Barnsley, from Messrs Cooper & Co. From this foothold they leased a further 900 acres of coal from the Edmunds Estate and proceeded to sink a colliery to the dip of the Silkstone Seam.

The first sod of the new undertaking, called Barrow Colliery, was cut on 4 June 1873. The new sinking was located some 450yds to the north east of their Worsborough Park Colliery acquisition. The sinking was a daring one since the deepest pit in the township at that time was Swaithe Main, at 240yds, and the Barrow sinking was planned to sink to the Silkstone Seam at 500yds.

The sinking contract was let to Barnsley Mining Engineers J.G. & A.Kell and sinking of three shafts proceeded rapidly despite serious setbacks. The two downcast shafts (Nos 1 and 2) were 15ft diameter and 43yds apart. The No.1 shaft was sunk to the Thorncliffe Thin Seam at

410yds, and the No.2 to the Silkstone Seam at 469yds. The No.1 shaft wound coal from the Parkgate and Thorncliffe Seams. Coal from the Parkgate Seam was 'dropped' to the Thorncliffe level, by means of a staple shaft, from where coal from the two seams was wound. The No.2 shaft wound coal from the Silkstone Seam.

The No.3 shaft was 17ft diameter and 58yds from the No.1 shaft. The No.1 and No.2 shafts were lined with bricks, tapered to the radius of the shaft, for a distance of 64.5yds from the surface. Beneath this were 30.5yds of cast iron tubbing, followed by brick lining to the shaft bottom. The No.3 shaft was lined with 68yds of brick from the surface, followed by 81yds of cast iron tubbing, then brick lined to the shaft bottom.

The Silkstone Seam was reached on 13 January 1876. This was celebrated by a banquet for 280 guests. The first coal was raised in September 1876 and within a few years the colliery was drawing 1,500tons per day and employed over a thousand men. The colliery was one of the most complete and best equipped in the country. It soon became a showpiece for the mining industry.

Pumping at the colliery was carried out from the Barnsley Seam, at a depth of 120yds, from one of the disused shafts of the Worsborough Park Colliery, using a Bull pumping-engine. At a later date the Bull engine was replaced by a geared horizontal engine which transmitted power to the pumps via 'L' legs at the shaft-top. The water pumped was used for coal washing, coke making and feeding the boilers.

The coking plant comprised thirty-two Coppee coke ovens and 122 oblong ovens. A further thirty Coppee ovens were added before the 1890s.

The contract for erection of colliery surface buildings and workmens' houses was let to Messrs Carruthers & Woodhouse, of Barrow-in-Furness. Bricks were provided by a patent brick-making machine, producing 11,000 bricks a day, and coal for the brickworks and steam engines was provided by Worsborough Park Colliery.

Barrow Colliery, 1975.

Worsborough Park Colliery closed a few years after Barrow Colliery opened. Coal was transported by a gravity operated incline plane to the canal-head at Worsborough. The new Barrow Colliery was linked to the Manchester, Sheffield & Lincolnshire Railway via a branch line from Dovecliffe station. A disadvantage was that main line locomotives could not enter the pit yard. Traffic from the pit had to be hauled to Dovecliffe, a quarter of a mile away, by colliery locomotives.

The company were eager to facilitate the movement of coal and coke to their iron and steelworks at Barrow-in-Furness. After an earlier, unsuccessful application to construct a line to join the Midland Railway, the company were eventually successful in getting a connection in 1898. A single line mineral branch line was constructed by the Midland Railway, to serve Barrow, Rockingham and Wharncliffe Silkstone collieries.

By the early 1890s the three lower seams, the Parkgate, Thorncliffe Thin and Silkstone were being worked, producing an output of 2,500tons per day. By the mid-1920s the colliery was working a large area of these seams on the north side of the colliery, and had just opened up the Fenton Seam. At the time the life of the colliery, in terms of the proved workable seams, was estimated as forty to fifty years. The Directors of the company realized that the colliery, which by then had been working for fifty years, needed modernizing in order to compete with the newer collieries to the east. New screening plant was erected at the surface. At the No.2 shaft, coal was formerly lowered down a staple shaft to the Thorncliffe Seam level, from where it was wound to the surface. Following the modernization a landing was provided to wind from the Parkgate level.

Barrow Colliery amalgamated with Barnsley Main and Monk Bretton Collieries in 1932, to form Barrow Barnsley Main Collieries Ltd, who owned the collieries until nationalization.

Developments at the colliery in the 1930s included: the replacement of the old timber headgears with steel headgears, construction of new pithead baths and a new canteen in 1934. The obsolete Coppee coke ovens were closed in 1936 and a new, larger battery was built as a separate undertaking and named the Barnsley District Coking Co.

Following nationalization in 1953, a major reconstruction of the colliery, at an estimated cost of £2,080,000 was approved, and finally completed in 1968. A new loading station was constructed at No.2 pit-bottom, improvements were made to the layout of No.1 pit-bottom, and return airways and locomotive man-riding roadways were enlarged. The reconstruction involved the driving of underground belt haulage roads, modification of the underground coal handling arrangements in order to handle two ton-capacity mine-cars, and erection of coal preparation plant on the surface.

In August 1962 the steam winding-engine at the No.3 shaft was replaced by an electric winder. In 1969 the two remaining steam-winders were replaced with second-hand electric winders. The No.2 shaft engine was replaced in May by an engine from Cannock. The No.1 shaft engine was replaced in August by an engine from Barnsley Main.

By the 1960s Barrow Colliery was at the forefront of face mechanization and this was reflected by a substantial increase in output, which almost doubled, to over 870,000tons, by 1962. During this period Barrow Colliery continued to break production records.

In the early 1960s Barrow featured regularly in the local newspapers. On 23 September 1961 the *Yorkshire Evening Post* reported:

Barnsley Pit Breaks County Output Record

One of the county's most mechanised pits.

The report went on to say the colliery had achieved a Yorkshire Divisional output record of 72.5cwt per man-shift.

On 14 October 1961 the Barnsley Chronicle reported: '**A European Mining Record?**' – noting that in the week ending 7 October, the Swallow Wood 9s face had produced an average face output of 22.53tons per man-shift. This was a production record for any 3ft thick seam in Britain, and probably Europe.

On Thursday 15 March 1962 the *Sheffield Telegraph* ran the following headline: '**The million tons a year show pit – BARROW – Pride of the North**.' The article reported that high-speed output, resulting from 100 per cent mechanization, had boosted production to such an extent that the colliery had earned the unofficial title 'show pit of the north'. The report went on to compare the output of 350,000tons in 1895, produced by pick and shovel from twenty-six coal faces, with 835,016tons from four highly mechanized faces in 1961. Productivity had risen to an annual record of 59.4cwt per man-shift for all workers and 212.6cwt for face workers alone. The article noted the reputation the colliery enjoyed for safety, indeed the colliery's motto read 'The motto of this pit is SPEED with SAFETY'. Mention was also made of the practice of underground stowage of dirt, into the worked out seams, reducing subsidence and surface spoil-heaps.

Barrow was also a pioneer in coal getting on a three-shift system. Roy Mason, Labour MP for Barnsley, said, 'If this system of mining can be successfully applied to more units in the coalfield it is the answer to the price war against oil.' A report in the *Daily Telegraph* of 11 May 1961 read:

> ***VALUE OF THREE PIT SHIFTS***
> *Mr John George, Parliamentary secretary, Ministry of Power, went underground yesterday at Barrow Colliery, Barnsley, to see the methods used in getting coal on a three-shift-system.*
> *The colliery is one of the pioneers in mechanization. Mr George said: 'We must have mechanization, and we must have it quickly in view of the serious competition with which we are faced.' He said: 'With a two-shift system we could just about survive. But with the three-shift system we should make a good profit.*

On 28 February 1968 the *Colliery Guardian* reported, '**Barrow goes for 1,000tons per day**'. The colliery was the first in the country to attempt to produce 1,000tons per day from one face on a regular basis. This was being attempted in the Thorncliffe Seam at the colliery.

In February 1971 a major scheme to link Barrow workings to Barnsley Main's No.2 shaft, for ventilation purposes, was started. Amendments to the original scheme were approved in April 1973 and included the construction of a ventilation drift from the Silkstone Seam, new main and standby ventilation fans, and a new methane extraction plant on the surface at the No.2 shaft. The Swallow Wood Seam was connected to the No.2 shaft in March 1973, and the drift to the Silkstone in November 1976.

In the early 1970s all three shafts were still in use. The No.1 and No.2 shafts, 372yds to the Parkgate, were used for winding men, coal and materials. The No.3 shaft, 371yds to the Parkgate, was used for winding men and materials from an inset at a depth of 308yds. Coal was produced from fully mechanized faces in the Swallow Wood, Parkgate, Lidgett and the recently opened Silkstone Seam. The total output was conveyed to a central loading station, then by mine-car to a common pit-bottom in the No.1 and No.2 shafts.

In this period retreat mining was being developed in the Lidgett Seam, the first time the new method of mining had been employed at the colliery, although it had been used successfully at other collieries in the Barnsley Area. Sixty percent of the output went to the colliery coking works, the remainder to power stations and industry generally. The coking plant closed in December 1975.

Barrow Colliery started to work coal in the Barnsley Main take in 1971-1972, with the last face in the original Barrow take finished working in June 1975. Details of this later phase in the history of the Barrow Colliery can be found in the section on the Barnsley Main Colliery.

Seams Worked:

Seam	Thickness (inches)	Depth (yds)
Swallow Wood Seam	36-64	147
Lidgett Seam	39	210
Flockton Seam	24	296
Fenton Seam	42	348
Parkgate Seam	54	372
Thorncliffe Thin Seam	39-48	408
Silkstone Seam	36-54	465

Brookhouse Beighton Colliery

Brookhouse Colliery was seven miles south-east of Sheffield on the exposed part of the coalfield. The coal measures at Brookhouse dip at a gradient of 1 in 12 to the east, and are intersected by a number of faults lying approximately north-west to south-east. The Spa and Whiston are the two major faults, which together resulted in a downthrow of the coal measures by 200yds.

The two shafts of the colliery were unique in that they were situated about three-quarters of a mile apart. The upcast Beighton shaft occupied the site of the much older Aston Colliery. Aston Colliery worked the Barnsley Seam from a pair of 12ft-diameter shafts, and ceased coal winding in 1886. In 1902 the Sheffield Coal Co. filled one of the shafts, widening the other to 18ft in diameter, and sank it to 436yds, slightly below the Silkstone Seam from which coal winding commenced in 1903. Ventilation for the pit was secured by a connection to Birley East Pit, some 1.5miles away. To assist the airflow a blowing fan was installed at Beighton Pit, but ventilation difficulties persisted, and in 1929 it was decided to sink a shaft at Brookhouse. Sinking of the Brookhouse downcast shaft began in March 1929 and the required depth of 441yds was achieved in October 1930. By 1931 a pit-bottom level had been formed at a depth of 420yds, some 20yds below the Thorncliffe Seam horizon, and connected to the workings by a drift. The shaft had a finished diameter of 20ft and was lined throughout with a monolith of concrete 9in thick.

The fan at Beighton was converted to an exhausting unit, the shaft becoming the upcast, while Brookhouse became the downcast for the colliery. The next stage was the development of the Thorncliffe Seam, as it had been decided that the Brookhouse shaft would deal with the output from both the Thorncliffe and Silkstone Seams.

A modern coal preparation plant, capable of handling 300tons per hour, was supplied and erected by the Coppee Co. (Great Britain) Ltd.

As a result of prevailing economic conditions, work at the colliery was suspended. It was not until 1937 that development was resumed when the Sheffield Coal Co. and the colliery were acquired by, and became a subsidiary company of, the United Steel Co. Ltd. The work under-

Beighton ventilating shaft, March 1984.

taken was to transfer the output from the Beighton Shaft to Brookhouse. A progressive policy was adopted and plans to equip the Brookhouse shaft on the most modern lines were begun. The programme was finalised by December 1941 and the changeover was completed by March 1942, when the whole of the output was wound at Brookhouse. The colliery was described as representing the last word in surface layouts, incorporating all that was best in modern mining practice. In the mid-1940s almost the whole of the output was won from the Thorncliffe Seam, and the Parkgate Seam was just being developed.

In the 1950s the colliery was working four seams; the Flockton at a depth of 310yds, the Parkgate at 375yds, the Thorncliffe at 400yds, and the Silkstone at a depth of 471yds, some 30yds below the bottom of the shaft. The seams provided coal for coking, gas production, locomotive and domestic uses. The coal produced by these seams was brought to a common pit-bottom, via cross measure drifts, for winding to the surface.

In January 1952 a Medical Treatment Centre was officially opened, and by the following April was on full time usage. In the late 1960s skip-winding was introduced into the colliery. Between 1974 and 1976 a project was carried out to develop the Silkstone Seam reserves between the Spa and Upper Whiston Faults, to replace diminishing reserves in the Thorncliffe Seam in the same area. The colliery was one of the casualties of the 1984-1985 miner's strike, closing in October 1985, soon after the strike finished.

Seams Worked:

Seam	**Depth (yards)**	**Thickness (inches)**
Flockton Seam	310	39-42
Parkgate Seam	375	42-60
Thorncliffe Seam	400	60
Silkstone Seam	471	36-48

Brookhouse shaft.

Darton Colliery

Darton Colliery was a small colliery situated about four miles north-west of Barnsley, and served by the Barnsley branch of the LMS. Two shafts were sunk at the colliery about 1914. The 12ft-diameter No.1 downcast shaft was sunk 89`yds deep to the Low Haigh Moor Seam, and the 9ft-diameter No.2 upcast shaft was sunk 126yds deep to the Lidgett Seam.

In the early 1940s the colliery was working the Top and Low Haigh Moor Seams. The Top Haigh Moor Seam was worked from short cross-measure drifts driven from the Low Haigh Moor Seam. Coal from both seams was wound from the Low Haigh Moor level in the No.1 downcast shaft. The Top Haigh Moor Seam was 2ft 9in thick and the full seam section was worked. Similarly the Low Haigh Moor Seam was 2ft 1in thick and the full section was worked. Both seams were 100 per cent machine cut and conveyed. In the early 1940s the weekly target output was 2,600tons. The colliery was owned by Fountain & Burnley at Nationalization in 1947, and closed in 1948.

Seams Worked:

Seam	Depth (yards)	Thickness (inches)
Top Haigh Moor Seam	80	33
Low Haigh Moor Seam	81.5	25

Dodworth Colliery

Dodworth Colliery, (alternatively known as Church Lane Colliery), was situated two miles to the west of Barnsley, near to Dodworth station on the Manchester, Sheffield & Lincolnshire Railway. In the early 1870s the colliery worked an area of coal greater than 1,000 acres, which was bounded on the western side by the outcrop of the Whinmoor Seam.

The colliery comprised three downcast coal drawing shafts and an upcast shaft. It was connected to the earlier Higham Colliery, about a mile distant to the north, which was also owned by Old Silkstone Collieries. Higham Colliery worked the Silkstone, Flockton & Parkgate Seams. At Higham there was a 10ft-diameter downcast Silkstone shaft, a 12ft-diameter upcast shaft, and a pumping shaft. To the Flockton Seam there was a 9ft-diameter shaft, and to the Parkgate a 10ft-diameter shaft. Higham Colliery became the pumping plant for Dodworth Colliery.

In 1857 two, 12ft-diameter, brick lined shafts were sunk, to the Silkstone Seam at 205yds, by Messrs Charlesworth, but were sold soon after to the Old Silkstone and Dodworth Coal & Iron Co., which was founded in 1862. One of the two shafts was an upcast and the other a downcast and coal drawing shaft. Close to the Silkstone upcast drawing shaft was the Flockton Pit. This was a 10ft-diameter, brick lined shaft, sunk 77yds to the Flockton Seam. In 1872-1873 a 12ft-diameter shaft was being sunk, close to the railway siding, to the Parkgate Seam at 120yds. The four shafts were located in close proximity to each other.

In the vicinity of the pits there were carpenters, blacksmith shops and stores as well as facilities for brickmaking. An excellent quality fireclay was raised from the pits. In the early 1870s preparations were being made for the erection of coke ovens.

Flooding from a neighbouring colliery, to the rise, resulted in flooding of the Silkstone Seam at Dodworth. In 1879 the colliery was closed and lay idle until 1899 when it was re-opened by the Old Silkstone Collieries Ltd who retained ownership until Nationalization in 1947.

By the early 1890s there were 110 beehive type coke ovens in operation, the waste gas from the ovens used to fire a number of the boilers in use at the colliery; three seams, the Silkstone, Parkgate and Flockton were being worked, with the Silkstone Seam producing the bulk of the output. By 1910 the three seams were together producing 1,400tons per day, a third of which was machine won.

In April 1907 a disastrous fire at 2.00 a.m. rapidly destroyed the whole of the surface plant including the three headgears. Fortunately the men working underground in the colliery at the time were all got out safely within twenty-four hours. New surface plant and headgears were erected and within fifty-two days the first of the new headgears was operational.

In 1910 Old Silkstone Collieries deepened the No.1 shaft at Dodworth to the Parkgate Seam. In 1912 coke ovens and a by-product plant were erected at the colliery. In 1924 the Silkstone Seam was de-watered by running the water into the shafts and the No.3 shaft was deepened a further 55yds to the Whinmoor Seam at 262yds. The sinking was undertaken to prove the seam, and then abandoned. The Silkstone Seam flooded again, the water level this time reaching some 40yds above the seam level.

On 30 July 1936 an Aerex axial flow fan was commissioned at Dodworth Colliery, by Walker Bros of Wigan. This was an early application of an axial flow fan in the Yorkshire coalfield.The Aerex was a departure from the 'old' centrifugal type of fan in use at collieries at that time, and had many advantages, not least of which was its simplicity. There had been substantial development at the colliery at the time, resulting in increased output, which prompted the need for improved ventilation.

The Flockton Seam was worked until 1943 when both the seam and drawing shaft were abandoned. In 1944 it was noted that the water level in the shafts was rising, indicating that the natural overflow to the pumping shaft at Higham was becoming impaired. At nationalization the output at the colliery was obtained from a concentration of working in three seams. Output was 1,200tons per day produced totally from hand filled faces. Ventilation had been improved and an electric winder installed at the Redbrook shaft.

In 1950 the decision was taken to reconstruct the surface of Dodworth Colliery and develop the Whinmoor Seam from the No.2 shaft. In order to develop the Whinmoor Seam, pumps were installed in the No.3 shaft and the shafts and workings to the rise in the Silkstone Seam. A major reconstruction scheme began in 1953 and was completed around 1958 at a cost of £2,180,000. The scheme involved the elimination of long underground rope haulages, the installation of a cable belt to convey coal to the surface via an existing drift, and the construction of a new coal washing plant. As a result of the scheme annual output rose from 336,000tons at the time of nationalization to 484,000tons in the mid-1950s.

In the 1950s the underground workings at the colliery were divided into three sections, Dodworth, Redbrooke and Rob Royd. Redbrooke and Rob Royd sending their coals underground to Dodworth via separate haulages. At Redbrook the Flockton and Fenton Seams were being worked, but the Flockton was nearing exhaustion, and it was planned to open out and work the Haigh Moor Seam together with a new area of the Fenton Seam. At

Dodworth Colliery, April 1982.

Rob Royd the Parkgate and Thorncliffe Seams were being worked, but the Parkgate had a very short life. In 1954 the Whinmoor Seam was being opened out, and developed for production in the Dodworth section.

In the period 1958-1962 the colliery was completely mechanized and the number of producing faces reduced from nine to five. The NCB spent over £100,000 on a surface reconstruction scheme, which included new pithead baths and a medical centre. Sport was also encouraged and a welfare scheme which cost about £30,000 provided football and cricket pitches as well as bowling greens and tennis courts.

In 1962 a surface drift was driven at the colliery and coal from the Whinmoor and Silkstone Seams was conveyed to the surface via the drift. In September 1965 the *Colliery Guardian* reported 'Redbrooke to be reclaimed'. The scheme was to pump about forty million gallons. of water from the workings into the River Dearne, which would enable some 6m tons to be extracted from a disused seam.

In the mid-1960s the situation for the colliery was not looking good and it was placed on the 'at risk list', but in August 1967 the Barnsley Area Director reported that 'The pit has made tremendous strides in the past six months', and as a result would remain open.

Higham and Redbrook Collieries were satellite sites to Dodworth Colliery and their output was conveyed overland to the coal preparation plant at Dodworth. Following the concentration scheme, Higham and Redbrook were closed as independent collieries and their coal resources worked from Dodworth.

In 1982 a new shaft was sunk at the Redbrook site for use as an upcast ventilation shaft, and the original Redbrook shaft, sunk in 1903, was used for man-riding. The Fenton Seam in the Redbrook section of the complex was abandoned about 1986, and at the same time the Thorncliffe Seam was newly developed. At the same time a new area of the Whinmoor Seam was due to start.

Dodworth Colliery had been scheduled for closure before the 1984-1985 strike, as part of the £400 million reconstruction of the Barnsley Area, which was carried out in the period 1979-1984. However the effect of the strike was to delay closure, and Dodworth men were transferred to the Redbrook section of the mine. Redbrook was given an expensive 'facelift' and fed its output into the West Side Complex centred on Woolley Colliery. With the merging of Dodworth and Redbrook in June 1987, the transfer of men to Redbrook and the raising of output at the Woolley site, Dodworth Colliery in effect closed. However. Redbrook did not survive for long, closing in December 1987 when Woolley Colliery closed.

Seams Worked:

Flockton Seam, 50yds deep, an excellent reputation as a house-coal
Fenton Seam, 90yds deep,
Parkgate Seam, 99yds deep
Silkstone Seam, 4ft 10in thick, 205yds deep,
Whinmoor Seam, 3ft 11in thick, 262yds deep

Elsecar Main Colliery

Elsecar Main was situated four and a half miles from Barnsley, mid-way between Barnsley and Rotherham, in an area of the South Yorkshire Coalfield which had been worked since the seventeenth century. The colliery take was approximately three miles by one and a half, and was generally free from major faults. The boundary of the colliery was formed by the 50yd depth of the seams from the surface, to the south-west, and Cortonwood and Manvers Main Colliery workings to the north and north-east respectively. The average dip of the seams in the take was 1 in 12 to the north-east. The colliery was served by a branch of the LNER.

The Barnsley Seam, which outcropped close to Elsecar, had been extensively worked in the district since the seventeenth century, and was worked out by the late nineteenth century.

Sinking of the shafts began in 1905, by the Earl Fitzwilliam Collieries Co., who owned the colliery until nationalization. The first sod for the No.1 shaft was cut on 17 July 1905, and for the No.2 shaft on 28 September 1906. The No.1 shaft first reached coal in the Parkgate Seam at a depth of 344yds, on 20 September 1906, and the No.2 shaft on 18 February 1908. The No.1 upcast shaft was 16ft diameter and 364yds deep. The No.2 downcast shaft was 18ft diameter and 370yds deep. The permanent steam winding-engines at the No.1 and No.2 shafts were installed in 1907. The No.1 engine worked until July 1960 when it was replaced by an electric winder.

A notable feature of the site was the colliery chimney, which was completed in October 1908. Standing to a height of 192ft, the chimney required almost 460,000 bricks in its construction. Following demolition in April 1985 the bricks were used to fill the colliery shafts.

Coal was first produced from the Parkgate Seam in 1908, and the seam was exhausted in 1961. Following its success in the Ruhr, trials of a Sampson Stripper (coal plough) began in the Parkgate Seam in September 1948. Following the exhaustion of the Parkgate Seam the

Elsecar Main Colliery.

main South Level Headings were maintained in order to ventilate and provide haulage for men and coal from the Thorncliffe and Silkstone Seams.

Royal visitors to the colliery included both King George V and King George VI. King George V and the Queen visited the colliery, and went underground, on 12 June 1912. King George VI and the Queen visited the colliery on 9 February 1944, but did not go underground.

In 1922 compressed air was introduced into the colliery to power Siskol coal-cutters in the Parkgate Seam. The employment of compressed air spread, and in 1923 compressed air engines were introduced and gradually replaced pit ponies for underground haulage. In 1927 the first electric coal-cutters were introduced into the Silkstone Seam. These were used in conjunction with compressed air shaker pans on the coalface. In 1924-25 the No.1 shaft was deepened to 534yds to enable the Silkstone Seam to be worked. The Silkstone was first worked in 1927 by a drift from the Parkgate Seam and was worked until it was exhausted in May 1975.
The Thorncliffe Seam, 20yds below the Silkstone, was reached by a drift from the Parkgate Seam and was first worked in 1938.

In the early 1940s the colliery was working the Parkgate, Silkstone and Thorncliffe Seams. The Parkgate Seam was hand got and filled into tubs, the Silkstone and Thorncliffe Seams were undercut with electrically powered undercutting machines, blown down and filled to conveyors.

The Haigh Moor Seam, some 200yds above the Parkgate, was opened out from headings driven from the shaft. Between 1942 and 1943 a surface drift was driven to the seam. First worked in 1945, output from the Haigh Moor Seam was conveyed to the surface by the 1 in 5 gradient haulage drift. The Haigh Moor Seam, was developed to replace output from the Parkgate Seam which was becoming exhausted. The Haigh Moor became the main source of production for the colliery due to the seam thickness, good working conditions and the high productivity achieved from working the seam by power loaded retreat mining. Productivity was further enhanced in 1956 by the installation of a cable belt down the drift and along the main South Level, a total distance of 2,630yds. In 1962 the cable belt was extended to a total length of 3,430yds. The drift proved to be an efficient transport system to the surface, and in conjunction with the output from the Haigh Moor Seam, helped to maintain the colliery's high profitability.

In 1950 the first diesel locomotives were taken underground. In the late 1950s and early 1960s a large-scale reconstruction of the surface and underground took place. This included a new coal preparation plant and improved underground transport. In July 1961 an electric winding-engine replaced the steam winder at the No.2 downcast shaft.

The output objective for the colliery for the financial year 1969-70 was a target of 845,500 saleable tons from the Haigh Moor, Thorncliffe and Silkstone Seams then in production. An estimate of the total reserves of the colliery at the time had been calculated as sufficient to provide a life of approximate five years working the seams in current production, and an ultimate life of twelve years if further workable reserves were developed in the Kent's Thick, Lidgett and Top Fenton Seams. The colliery was producing coal for coke making, the gas, domestic and power station markets. The plan was to work the Haigh Moor Seam until March 1973, and to develop the Lidgett and Kent's Thick Seams, prior to this date, to replace Haigh Moor, Thorncliffe and Silkstone Seam output.

Coal from the Haigh Moor Seam was transported, by belt conveyor, to the 745yd-long, 1 in 5 gradient drift, to the surface and direct to the coal preparation plant, a total distance of some 3,500yds. The belt carried 444,067tons in 1957, increasing to 664,000tons in 1967. The

No.1 shaft was used for man-riding to the working seams, and for materials and dirt-winding from the Haigh Moor Seam. In 1967 the 1,500 men at the colliery achieved its record output of 1.13m tons, an output of over 1m tons for the first time in the colliery's history.

In 1971 the 1 in 5 gradient drift was extended from the Swallow Wood to the Lidgett Seam, which replaced the Thorncliffe Seam, which was exhausted by November 1972. The Lidgett Seam was first worked in December 1972, but the seam proved difficult to work and production was poor.

At the end of April 1974 a coalface at the colliery was abandoned in order to safeguard historic buildings in the nearby village of Wentworth. In January 1979 the Kents Seam was developed and its coal conveyed to the surface via the drift. This seam also proved difficult to work, with water the principal problem. On Friday 5 June 1981 the last train of coal left Elsecar Main Colliery. This comprised household cobbles bound principally for the Channel Islands, via the port of Goole. The colliery finally closed in October 1983, just prior to the miners strike of 1984-1985, but stood during the strike and was not demolished until the strike finished.

Seams Worked:

Seam	**Worked**	**Depth (yards)**	**Thickness (inches)**
Kent's Thick Seam	1979-1981	30	43
Swallow Wood Seam	1945-1983	152	52
Lidgett Seam		199	26
Parkgate Seam	1908-1969	344	54
Silkstone Seam	1927-1975	450	33
Thorncliffe Seam	1938-1972	470	24

Grange Colliery

Originally sunk in 1845 to work the Silkstone Seam, purchased from G.W.Chambers & Sons in 1890, Grange Colliery was situated three miles from Thorncliffe, and was served by the LNER. The colliery had a 9ft-diameter downcast shaft and a 7ft-diameter upcast shaft, sunk to the Silkstone Seam at 53yds.

In the mid-1890s an output of 150tons per day was produced and sixty beehive ovens were at work producing coke. A haulage engine, placed at the surface, operated an endless rope haulage system down the shaft and into the east dip haulage road. An air-compressor at the surface supplied compressed air down the shaft and along the east dip haulage road, to operate ram pumps forcing water to the pit-bottom. From the pit-bottom the water was pumped to the surface.

In 1921 the colliery was closed and the entire plant scrapped. New plant was installed, and by mid-1927 the colliery had re-opened, with the new screening plant almost complete. The small diameter shafts had a limited capacity of some 400tons per day, and it was decided to drive a drift at a gradient of 1 in 6 to intercept the Silkstone Seam. The drift was driven in the mid-1920s, and was driven from the surface for 790yds. By the late 1920s output had increased to 500tons per day, and was planned to increase to 1,000tons per day.

In the mid-1920s a drift was driven at Hudson's Rough to serve as an upcast airway and as a travelling way for the men working in the Silkstone Seam. The fan at Hudson's Rough was the first Stewart-Walker axial type fan installed at a colliery in the UK, and was commissioned in June

1928. The fan had fourteen, 10ft-diameter, twin-blade laminated wooden aerofoil stages, similar to aircraft propellers. The blades were mounted on a shaft and acted like a large spiral screw, rotating at high speed in a cylindrical tunnel connected to the main drift. The fan proved to be cheap and efficient, and could be easily reversed to comply with the Coal Mines Act of 1911.

In 1922 a 1 in 5 drift, known as Bradgate No.1 Drift, was put down to the Silkstone Seam from a quarry at Bradgate. This formed a second outlet for Grange Colliery and served as a service pit to provide landsale coal for Rotherham. In about 1925 a second drift was put down at the quarry and became known as Bradgate No.2 Drift, and in 1925 a third drift was driven from Hudson's Rough to the Grange workings. This drift became the upcast for the colliery and was extended in order to convey the men into the workings by 'paddy' mail. Bradgate Drift was also a pumping station.

The workings were drained by pumps at Bradgate Drift and Hudson's Rough, and the rise workings to the north, by the pumps at Thorpe Hesley Pit. A statement of pumping, dated 1935, showed the average quantity pumped per day in summer months and the maximum capacity of pumping plant (in brackets).

Bradgate	108,000gallons (432,000gallons)
Hudson's Rough	13,000gallons (101,000gallons)

In the late 1940s the colliery was working the Silkstone Seam and the Thin Coal, with approximately three-quarters of the output produced by the Thin Coal. The decision to close the colliery had been taken during 1961, and on 21 April 1962 the *Sheffield Telegraph* reported:

COLLIERY TO BE CLOSED

Worked out by end of year

Grange Colliery, Rotherham, a series of drifts and shafts, is to close down this year, an NCB spokesman said yesterday. The colliery will be worked out before the end of the year.

'Work will be carried on in the pit until coal deposits are exhausted… There will be no redundancy. Men will be transferred to other collieries in the area as and when their services are no longer required at the Grange Colliery.'

…Since late last year about 100 men have left the pit… about 140 men still worked there. 'We have been told everyone will be transferred to other pits before July."

Seams Worked:
- Thin Coal-seam
- Silkstone Seam

Haigh Colliery

In the early 1940s the colliery employed surface drifts to access the Top Beamshaw and Top Barnsley Seams. The drifts in the Top Barnsley Seam surfaced close to Haigh railway station. The colliery was served by the Barnsley branch of the LMS.

A 12ft-diameter shaft at Bolton Wife Hill, near Crigglestone, sunk 65yds to the Low Beamshaw, was employed as the upcast shaft for the Top Beamshaw Seam. The Winter Seam, which lay 22yds above the Top Beamshaw, was accessed by a 1 in 6 gradient cross-measures drift from the Top Beamshaw. Both the Winter and Top Beamshaw Seams suffered from water and were wet to work in. The Top Barnsley was one hundred per cent machine cut and conveyed by the early 1940s. The full seam section of the Winter and Top Beamshaw Seams was worked, but only 39in of the Top Barnsley Seam section. The colliery was owned by Fountain & Burnley at nationalization in 1947.

In the late 1950s the colliery was connected to Woolley Colliery by a long conveying scheme, using trunk conveyors to transport the coal from Haigh to the new washery at Woolley Colliery. On 9 August 1968 Haigh Colliery worked its last production shift and the colliery closed due to exhaustion of its coal reserves.

Seams Worked:

Seam	Depth (yards)	Thickness (inches)
Winter Seam	*c.*38	36
Top Beamshaw Seam	*c.*60	32
Top Barnsley Seam	*c.*150	72 (including a 13in dirt parting)

Handsworth Colliery

Handsworth Colliery was situated approximately two miles from the centre of Sheffield. Sunk by the Nunnery Colliery Co., who retained ownership until nationalization, the colliery was sunk with the objective of extending the limits of Nunnery Colliery, and was often referred to as Handsworth Nunnery Colliery. The shafts were ultimately sunk between 1901-3, to a little over 409yds, intersecting the Silkstone Seam at a depth of 439yds.

In the late 1920s a total of twenty-five Becker regenerative coke ovens were erected near the Handsworth site. This was the first application of American coke oven practice to British conditions. The plant produced coke, tar, sulphate of ammonia, naphthalene and supplied gas to the Sheffield Gas Co. The coking and by-products plant was situated about a third of a mile to the north of the colliery, alongside the LNER line, at a point near to the Waverley Drift mine.

In 1950 the working of the Haigh Moor Seam commenced via a surface drift, coal winding in the shafts ceased at this time. Handsworth coal was then washed at other collieries in the area. Mechanization had been unsuccessfully tried at the colliery, but in February 1965, success was achieved. Despite the benefits of mechanization and the results achieved, the reserves of coal at the colliery were limited. The colliery suffered from poor geological conditions, wet workings and had reserves estimated sufficient for two years.

In November 1965 it was announced that the colliery was scheduled for closure. Despite the announcement of closure, output performance at the colliery continued to increase. This was attributed to good labour relation, high morale and the effect of mechanization. The 470 men at the colliery at the time were fully consulted about the effects of the closure and were assured of employment, in their teams, at neighbouring collieries.

In October 1966 the last colliery within the boundary of the city of Sheffield ceased production due to the exhaustion of its reserves, and finally closed one year later in October 1967.

Seams Worked:

Seam	Depth (yards)	Thickness (inches)
Haigh Moor Seam	135	53
Flockton Seam	259	64
Parkgate Seam	318	63
Silkstone Seam	439	65

Hoyland Silkstone Colliery

Hoyland Silkstone Colliery was situated on the edge of the village of Hoyland, four miles south-east of Barnsley, close to the Midland Railway. The mineral field associated with the colliery comprised about 1,600 acres, owned principally by Earl Fitzwilliam and T.F.C. Vernon-Wentworth of Wentworth Castle, near Stainbrough. The mineral property was leased by the Hoyland Silkstone Coal & Coke Co. Ltd in 1873.

The colliery had three shafts, the 8ft 6in diameter Flockton Pit, the 22ft diameter Silkstone Pit, and the 14ft diameter Thorncliffe Pit. The Flockton Pit was sunk to the Barnsley Bed, Haigh Moor and Lidgett Seams, and the Thorncliffe Pit to the Silkstone Seam. The Silkstone and Flockton were downcast shafts, and the Thorncliffe an upcast shaft.

Sinking of the Silkstone shaft was entrusted to John Higson, a well-known mining engineer of Manchester, and Mr W.H. Peacock, who was the resident viewer of Hoyland Colliery. Sinking of the shaft commenced in March 1874 and proceeded rapidly, reaching

Hoyland Silkstone Colliery, early 1900s.

the Silkstone Seam at a depth of 510yds on 7 January 1876. Early fears that the Silkstone Seam, at such a great depth, would not be so thick as at pits where the coal was at shallower depth, soon proved to be wrong. At Hoyland, the Silkstone proved to be 5ft 8in thick and of excellent quality. The shaft was 20ft in diameter and brick lined from top to bottom. When it was sunk the Silkstone Pit was believed to be the finest in the country, being twice the depth of the only other shaft of the same diameter.

Coal was drawn in all three shafts; Silkstone coal from the Silkstone Pit, Thorncliffe and Parkgate coal from the Thorncliffe Pit, and Flockton coal from the Flockton Pit. The Barnsley Bed was already exhausted when the Silkstone shaft was sunk in 1873, having been worked-out by a company who leased a portion of the royalty prior to sinking.

At one time the Silkstone Seam had been worked too close to the shaft pillar causing movement in the pit-bottom. Girders from the Tay Bridge (after the disaster) were brought to Hoyland, and used to reinforce the pit-bottom. This, however, did not prevent the movement, which bent and crushed the beams.

The principal seams of coal intersected in the shafts were as follows:

Seam	**Depth (yds)**
Barnsley Bed	135
Flockton	329
Haighmoor	196
Lidgett	256
Parkgate	408
Thorncliffe	432
Silkstone	509

Coal was conveyed from the colliery via the Midland and Great Central Railways and on the Aire & Calder Canal, to which the colliery was connected by a mile-long branch line. At the close of the nineteenth century coal from the colliery was distributed throughout Yorkshire, the south of England and London. Hoyland Silkstone coal was shipped from the ports of Hull and Goole.

In the late nineteenth century a daily output of 800tons per day was wound from the Silkstone Seam at 509yds. There were 108 12ft-diameter, beehive type coke ovens arranged in three double rows at the colliery. Waste heat from the ovens was used to heat boilers at the colliery.

In the late 1890s steam powered compressed air plant was installed. The intention was to use compressed air-powered coal-cutting machines to work the thinner seams at the colliery. In February 1917 a new coke oven plant was started. The new plant comprising a Luhrig washer, thirty-seven Semet-Solvay ovens, power plant and auxiliaries. It was constructed for the colliery by the Coke Oven Construction Co. Ltd of Sheffield. The plant dealt with all the small-coal, nuts, peas and coking slack produced at the colliery. Waste flue-gases supplied heat to Lancashire boilers. Shale and refuse from the coking plant was transported away to the spoil heap by aerial ropeway.

On 1 April 1925 Newton, Chambers & Co. Ltd purchased Hoyland Silkstone colliery, which they closed in 1928, working the reserves from Rockingham. Hoyland Silkstone then became a service pit for Rockingham. A workman at Hoyland Silkstone was awarded the Albert Medal for bravery.

Seams Worked:

Flockton Seam, 2ft thick, which furnished a good quality housecoal.

Haigh Moor Seam, yielded 3ft 9in of coal separated by 5ft of mudstone.

Lidgett Seam, which yielded a total of 3ft 1in of coal.

Parkgate Seam, 3ft 5in thick (excluding a 3in thick stone parting). The upper Hards section yielded a steam-coal and the lower Softs section a gas-producing quality coal.

Thorncliffe Seam, varying from 1ft 8in to 2ft 4in in thickness.

Silkstone Seam, a total of 3ft 10in thickness, comprising: Hards, 9in thick, Top Softs, 1ft 1in thick, and the Bottom Softs, 2ft thick. The Hards yielded the best house-coal in the district and the top and bottom softs an excellent quality gas-coal.

New Stubbin Colliery

New Stubbin Colliery was sunk approximately two and a half miles north of Rotherham, close to the village of Rawmarsh, by Earl Fitzwilliam Collieries Ltd, who retained ownership of the colliery until nationalization. Situated in a narrow, somewhat constricted, valley the colliery was served by the LMR and the LNER.

At the beginning of the twentieth century, with the realization that the reserves of the Top Stubbin Pit would soon run out, Earl Fitzwilliam prepared to sink a replacement colliery. The Earl had been advised by his agents to sink a pit at Cantley, in the Doncaster district, but chose to sink in the immediate vicinity so that his workmen could continue their association with the colliery company in their own area. Preliminary surveys were carried out during 1910, and it was decided to sink the pit adjacent to the Earl's Bank Pit Brickworks.

The first grass sods were removed from the new site on 13 November 1913 at the No.1 shaft, by Viscount Milton, the two-year-old son of Earl Fitzwilliam, and at the No.2 shaft on 16

New Stubbin Colliery, 1980.

September 1913. Construction of the colliery proceeded rapidly, using bricks from the Earl's brickworks. Sinking of the shafts commenced from the bricking cribs, at the No.2 shaft on the 13 May 1915 and at the No.1 shaft on 1 June 1915. The 6ft thick Parkgate Seam was reached, at a depth of 287yds, on 13 January 1916 in the No.2 shaft and on 27 April in the No.1 shaft. The two shafts were connected by airways in the Parkgate Seam on 19 May 1916. New Stubbin was one of the few English collieries to be sunk during the First World War. The average speed of sinking for the two shafts was recorded as nine to 10yds per week. The No.1, downcast, shaft was 18ft in diameter and the No.2, upcast, shaft 16ft in diameter. Shaft sumps were constructed in both shafts to a depth of approximately 18yds below the Parkgate horizon. The first coal was wound out of the No.1 shaft, from the Parkgate Seam, on 13 November 1919 and serious coal production commenced in 1920. By this time the reserves at the Earl's Top Stubbin Pit were exhausted and men from the pit were transferred to the new sinking.

When the colliery was first opened the coal was won by hand, each man working his own stall (piece of work) along a short advancing face. This system of working continued until the first hand-got longwall face was established in 1921. Soon after Siskol coal-cutters were used to undercut the coal, and in the late 1920s Hopkinson type coal-cutters were introduced.

Coal from the colliery was transported down the Stubbin Incline, the Earl's private line, to the junction with the Great Central & Midland Railway, at Parkgate. Coal was processed by the large by-product plant at Parkgate, and the colliery continued to supply the South Yorkshire Chemical Co.'s works at Parkgate for many years.

On 29 November 1930 the Miner's Welfare Scheme opened the first baths at the colliery, under the Mining Industries Act of 1926. In 1935 the colliery produced the highest weekly output in its history, at 16,189tons, a figure unbeaten even in later years when the latest mining technology was employed.

During 1935 a drift was driven from the Parkgate Seam to the Thorncliffe Seam, which lay 27yds below the Parkgate. Production in the Thorncliffe commenced later in 1935, its coal coming up the drift to be wound at the Parkgate level. Production in the Thorncliffe Seam continued until abandonment on 20 April 1973.

In the early 1940s both the Parkgate and Thorncliffe Seams were being worked. Some 5ft of the full section of 5ft 8in of the Parkgate Seam, and the full section of 2ft 8in of the Thorncliffe Seam, were worked. The Thorncliffe Seam was connected to the Parkgate level by a pair of drifts, one at a gradient of 1 in 6 and the other at 1 in 2. Coal from the Thorncliffe Seam was wound up the No.1 downcast shaft, though both shafts were equipped for coal winding. The general dip of the seams was 1 in 12 to the north, but increased to 1 in 5 to the south making working conditions more difficult in that area. On 5 February 1952, the *Yorkshire Evening Post* reported: 'NCB approve £985,000 development plan to yield extra 270,000tons in the Silkstone in the next 3 years.'

A major reconstruction commenced in October 1951 and was completed in 1957. Due to rapid exhaustion of the Parkgate Seam it was necessary to develop workings in the Silkstone Seam. The preparation to work the Silkstone entailed a considerable amount of development and expenditure, and it was decided to fully mechanize the colliery, both underground and at the surface. This involved deepening both shafts, constructing a new pit-bottom at a depth of 422yds, electrifying the winding-engines, modernizing the surface facilities and provision of a new coal preparation plant, at a total expenditure of £1.25million.

Shaft deepening from the Thorncliffe to the Silkstone was carried out by the Blandford-Gee Cementation Co. Deepening of the No.2 shaft commenced in July 1951 and was completed in October 1952. Deepening of the No.1 shaft commenced in November 1952 and was completed in March 1954, and a new pit-bottom constructed to facilitate the use of diesel locomotives. The shafts were deepened from the Parkgate to a level some 50yds beneath the 371yd deep Silkstone Seam. During shaft deepening the last of the pit ponies working in the Parkgate Seam were brought out of the pit. At the same time as the shafts were being deepened, a development face was being worked from Elsecar Main towards New Stubbin, in order to develop the Silkstone and to link up with two locomotive drifts driven from the new pit-bottom at New Stubbin. The stone drifts were driven at gradients of 1 in 200 and 1 in 300. A pair of staple shafts were sunk between the shafts, so that all of the coal mined could be wound from the new horizon. No.1, a 12ft-diameter and 31yd deep staple shaft, was sunk between January and August 1954, from the Parkgate to the Thorncliffe. No.2, a 12ft diameter and 54yds deep staple shaft from the Thorncliffe to the Silkstone, was sunk between September 1954 and June 1955. These enabled all the coal to be transported, in two ton-capacity mine-cars, by diesel locomotive to the new pit-bottom.

The No.2 winder was converted from steam to electric during the annual shutdown between 25 July and 7 August 1954, and the No.1 winder two years later between 28 July and 12 August 1956. The No.1 winder wound coal, the No.2 winder wound men, dirt and materials.

Construction of a new coal preparation plant was commenced in January 1954 and was in operation by March 1955. Production from the Silkstone commenced in 1955, the first Silkstone coal going to the new washery on 13 September. Prior to reconstruction of the colliery, and installation of the new coal processing plant, no dirt was wound out of the pit, all dirt was stowed underground in the Parkgate Seam. Commissioning of the washery and working of the Silkstone Seam saw the formation of the first spoil heap at the colliery.

As a result of the modernization programme, output was raised from 437,000tons in 1950, to 532,000tons in 1958. In an NCB letter, dated December 1953, the estimates of the reserves remaining were, Parkgate 4m tons, Thorncliffe 9m tons, and Silkstone 11m tons. Of this total some 16m tons were accessible. Based on these figures it was estimated that the colliery had a life of thirty years.

Mechanization and power loading, using British Jeffrey Diamond coal-cutters and coal conveyor system, was introduced into the Thorncliffe Seam in 1957. Development of the south area of the Silkstone commenced in 1959, and the thinning of the seam saw the introduction of a plough face and new hopes of working thin seam coal safely and effectively. In the mid-1960s Trepanner shearers were introduced into the Thorncliffe Seam.

On 1 January 1959 the *Colliery Guardian* reported that the colliery was experiencing water seepage into its workings, which threatened to flood the colliery. The water was seeping from a disused drift mine close to the Thorncliffe Seam. Pumping was taking place around-the-clock, and was also taking place at Grange Colliery. The NCB reported that the pumping was adequate and said that there was no danger of flooding.

Closure of the nearby Aldwarke Main Colliery during 1961 resulted in the transfer of men to a number of collieries in the area, including New Stubbin.

On 15 June 1964 a man-riding 'ski lift' was commissioned, in a drift driven between the Silkstone and Thorncliffe, one of the earliest installations of its kind to be used underground. The 'ski lift' operated in a 1 in 3.2 gradient drift.

In 1966 the Parkgate was finally exhausted and coal working ceased on 18 December, after some forty-six years. In 1973 the workable reserves in the Thorncliffe were also exhausted, focusing attention on the remaining reserves in the south area of the Silkstone, where the seam thinned to 26in of coal. Despite the introduction of heavy-duty German Gleithobel coal ploughs, the remaining thinned area of Silkstone proved to be unworkable, and on 6 July 1978 production in the Silkstone ceased and the colliery closed, having produced 21m tons of coal.

Seams Worked:

Seam	**Depth (yards)**	**Thickness (inches)**
Parkgate Seam	287	72
Thorncliffe Seam	317	36
Silkstone Seam	371	30

North Gawber Colliery

North Gawber Colliery was situated in the Parish of Darton, near to Mapplewell and Staincross, some two and a half miles north of Barnsley, and six and a half miles south of Wakefield. The colliery was connected to the Barnsley-Wakefield Branch of the Lancashire & Yorkshire Railway, near Darton station, by means of a one and a quarter mile private mineral line.

The colliery was situated in an area where a number of seams outcropped at the surface, and had been worked over a long period of time. References to the mining of coal in the area are made in court records dating back to the Middle Ages, the earliest reference dating from about

North Gawber Colliery, 1981.

1387. The Beamshaw, Kent's Thick and Kent's Thin were the principal seams, among seven that outcrop in the area, and the three seams outcropped on the colliery premises.

Following the opening of the Lancashire & Yorkshire Railway branch line in 1850, mining development in this area of South Yorkshire intensified, and it is from this period that the colliery of North Gawber dates.

The Thorpe family of Gawber Hall commenced the sinking of North Gawber Colliery in 1850. Messrs R.C. Thorpe & Co. sank two shafts, reaching the Barnsley Seam, at a depth of 103yds, in September 1852. The shafts comprised the No.1, a 9ft-diameter downcast and drawing shaft, and the No.2, a 14ft diameter upcast shaft located 80yds north-west of the downcast shaft. In addition a 9ft diameter water shaft was sunk 80yds east of the No.1 shaft.

The Barnsley Bed was of high quality and some 10ft in thickness. As in the case of nearby Woolley Colliery, the Barnsley Seam was separated, by 6-10in of dirt, into the 'Softs' and 'Hards', at the shafts, which increased in a north-easterly direction to more than 20ft. This resulted in a division of the Barnsley into two distinctive seams, the Barnsley Softs and Barnsley Hards.

The colliery mineral take comprised 3,400 acres, with the seams dipping, at two and a half inches to the yard, towards the north-east. It was remarkable in that it was almost totally surrounded by major faults, which determined the boundaries of the colliery.

In 1858 a beam pumping engine was erected on the water shaft. The engine was named after John Clarkson Sutcliffe, who was manager at Gawber Hall and North Gawber Collieries for over forty-five years, until his death at the age of seventy in 1858. The engine was still in regular service until its demolition in 1962, after which the pumping shaft was filled. A drift entrance to the mine, below Spark Lane, near the mineral line, was used as an access to the mine by both men and ponies.

In 1872, during a period of prosperity in the coal industry, the two pits owned by Messrs Thorpe, North Gawber and Willow Bank, became a joint stock company with a capital of £150,000. However, in the late 1870s the colliery suffered a slump in trade and the owners got into financial difficulties. In 1880 the two Thorpe pits were offered for sale as going concerns. North Gawber Colliery stood idle from 1880 until 1882. During this time only the coking plant was operating, the remaining colliery plant kept on maintenance. Following this slump Willow Bank Colliery closed and North Gawber was taken over by Messrs Fountain & Burnley, who owned the colliery from 1882 until nationalization. Fountain & Burnley also owned the nearby Woolley Colliery.

By the 1890s there were eighty-two beehive type coke ovens in operation at the colliery, producing about 550tons of fine furnace coal, and the screens, pit bank and pit-bottom were illuminated by gas produced at the plant.

In 1926 an 18ft shaft was sunk to the Lidgett Seam at a depth of 217yds, intersecting the Top Haigh Moor Seam at a depth of 171yds and the Low Haigh Moor at about 177yds. Production from the Lidgett commenced in 1930. All three seams were worked from the shaft. This shaft, the No.3, became the downcast for the colliery, with one of the old Barnsley shafts retained as the upcast. The Barnsley Seam was worked continuously at the colliery until 1934, when its output was replaced by that from the Haigh Moor Seam.

From 1934 there were two coal winding shafts in operation, the Barnsley shaft drawing from the Haigh Moor Seam, and the new No.3 shaft drawing from the Lidgett Seam. This system continued in operation until 1942, when a scheme for the concentration of haulage and winding was completed, and output from the colliery was concentrated at the new No.3 shaft.

From 1930 onwards a gradual reconstruction of the colliery took place. In 1938 a new lamp room and pithead baths, capable of accommodating 1,000 men, were erected. In 1948 a canteen was constructed. A 200ton-per-hour Baum type washer, one of the first to be ordered by the NCB commenced operation in March 1950. A new ventilating fan, an axial flow Aeroto type, was installed in 1952. An aerial ropeway, commissioned in April 1941, was used to transport between seventy and seventy-five tons of dirt per hour to the spoil tip.

In the early 1940s the colliery was working the Lidgett, and Top and Low Haigh Moor Seams, both were 100 per cent machine cut and conveyed. The principal difficulties were faults in the Top Haigh Moor, and poor roof conditions in the Low Haigh Moor. Lidgett and Low Haigh Moor coal was wound from the Lidgett level in the No.3 downcast shaft, Low Haigh Moor coal transported down a cross-measure drift sloping at 1 in 1 to the Lidgett horizon. Top Haigh Moor coal was wound up the No.2 downcast shaft from the Barnsley level, transported up a 1 in 3 gradient cross-measures drift to the Barnsley level.

In August 1948 an underground concentration scheme was completed and the output from the adjoining Darton Hall Colliery, which worked the same three seams, the Top and Low Haigh Moor and Lidgett Seams, was wound at North Gawber No.3 Lidgett shaft. This increased the output wound at North Gawber from 7,600 to 10,400tons per week. Following the concentration scheme the manpower at Darton Hall was transfered to North Gawber, the Darton Hall shafts remaining in use for ventilation and pumping until 1970, when they were filled and capped.

Underground reconstruction and extensive mechanization kept pace with the improvements at the surface. A particular feature of the colliery was the extensive use of underground roadway lighting, with a total of 9,000yds of main roadway lit by the early 1950s. By 1950 the colliery had an output of some 2,500tons per day and was completely electrified.

In 1957 a scheme was initiated to develop the upper coal-seams. The scheme entailed driving a pair of drifts from the surface to the Beamshaw and Winter Seams. Development of the Winter, Beamshaw and Kent's Thick Seams was required to replace output from the Lidgett Seam, which was nearing exhaustion.

In 1961 a face opened in the Beamshaw Seam proved to be unworkable due to poor roof conditions and excessive water. In 1962 the face was abandoned. In 1965-66 a further drift was driven, extending from the existing surface drifts, to access the Kent's Thin and Top Haigh Moor Seams. The new underground drift connected into the No.1 intake drift, of the original pair of drifts. This new drift enabled the whole of the output from the lower seams to be transported up the mile long drift, rising 1,000ft in the journey, into the No.1 intake drift and to the surface. This enabled coal winding at the shafts to be dispensed with. On completion of the scheme the Nos 1 and 2 shafts were filled.

In May 1971 the *Colliery Guardian* reported a £76,000 development to access 2.5m tons of reserves in the Kents Thick Seam. To access the reserves a 400yd-long drift was driven into the seam. The new workings would gradually replace output from the Top and Low Haigh Moor Seams. The first face was planned for production in December 1971.

The number of Pit ponies had been gradually reduced as the colliery became more mechanized, and were finally retired as a means of underground haulage in 1972.

In the period 1979-1984 nearby Woolley Colliery became the hub of the £102 million 'West Side' complex of the Barnsley Area. The central coal preparation plant at Woolley received coal from a number of collieries, including North Gawber, which merged with Woolley in January 1986. North Gawber closed in December 1987, when Woolley Colliery closed.

Seams Worked:

Seam	**Depth (yards)**	**Thickness (inches)**
Winter Seam		
Beamshaw Seam		
Kents Thick Seam	135	36
Barnsley Seam	102	120
Top Haigh Moor Seam	173.5	34
Low Haigh Moor Seam	179	28
Lidgett Seam	217	32

Nunnery Colliery

Located close to the centre of Sheffield, the colliery was one and a half miles to the east of the parish church of Sheffield, on an elevated tract of land. The colliery railway sidings connected to both the Midland Railway and the Manchester, Sheffield & Lincolnshire Railway systems. The company also had a number of privately owned lines, which gave access to various coal depots and coke ovens, which were to be built at the Handsworth site in the 1920s.

The colliery was sunk by the Duke of Norfolk, on his estate, between 1860-64. Two shafts were sunk to the Silkstone Seam at a depth of 235yds. The downcast shaft, later called the Old Silkstone Shaft, was 15ft 6in in diameter and divided into two sections by a timber brattice. The smaller shaft section was occupied by pumps and the larger section was used to raise Silkstone coal, which was raised in the shaft until 1896. The upcast shaft, 88yds south-east of the Old Silkstone shaft, was 12ft 6in in diameter, and was used to raise Parkgate Seam coal from a depth of 115yds.

In 1874 a group of capitalists in Sheffield, acquired the lease for the colliery, and subsequently worked it under the title the Nunnery Colliery Co. The company then went on to run the undertaking until nationalization in 1947. A feature of considerable consequence to the mining operations at the colliery, was the occurrence of a major fault, at a distance of 132yds east of the two shafts originally sunk to the Silkstone Seam. In order to work the large area of Silkstone coal to the dip of this fault it was necessary to sink a new shaft.

In 1892, electricity was introduced into the colliery, when a 500V DC generator was installed. The generator was installed primarily for the purpose of pumping from the south-east district in the Silkstone Seam. The installation ran for eleven years until it was taken out in 1903. In 1910 the use of electricity was again introduced and its use gradually extended. Electricity was used to power underground haulage. By 1924 the aggregate brake horsepower of electric motors in use totalled 1,385, of which some 380 were employed underground.

In 1893 the New Silkstone Shaft was commenced, and sunk to a depth of 315yds, sinking taking one year. The new shaft was situated about 200yds north-east of the Old Silkstone Shaft, beyond a fault which resulted in a downthrow of the seams by 68yds. The new shaft was 14ft in diameter and cut the Parkgate at a depth of 205yds and the Silkstone at 315yds. The new shaft enabled Silkstone coal to be worked beyond the fault. By the late 1890s all Silkstone

Nunnery Colliery, early 1900s.

coal was raised in this shaft. Silkstone coal worked by the original shafts was by then exhausted. In the 1890s a stone drift was driven from the Old Silkstone shaft to meet the Parkgate Seam beyond the fault. By the early 1900s Parkgate coal was being raised in this shaft.

In 1893 an 11ft diameter staple shaft was sunk between the Parkgate and Silkstone Seams, in the area beyond the fault, to act as a return airway. When the remaining pillars of Parkgate coal, in the upper area, were worked out in the late 1890s coal winding in the Parkgate upcast shaft ceased. The shaft then reverted to man-riding for the men employed in the Parkgate Seam, and as the upcast for the colliery. At this time there were 145 beehive type coke ovens at the colliery, producing 200tons of coke per day, which was in high demand for steel-making.

The Silkstone Seam produced a particularly clean-burning coal well suited to household use. A major market for the Silkstone coal produced by the colliery was for domestic use in Sheffield, where over half of the house-coal used derived from Nunnery Collieries. A speciality of the collieries was the manufacture of the finest steel-melting coke, of which about 200tons per day were produced, and for which there was a large demand amongst the crucible steel manufacturers in the city. The Parkgate Seam furnished an excellent steam-coal, which was also mainly used locally. In the late 1890s the colliery was producing 800tons of Silkstone and 900tons of Parkgate coal per day.

The fan erected at the No.2 shaft by Walker Bros of Wigan, in 1891, was driven by a tandem compound steam engine, an early application of this type of engine for driving ventilating fans.

The colliery was well furnished and had one of the largest and most powerful pumping plants in the country, comprising a double-acting condensing beam pumping engine. The engine was worked day and night, speed varying with the season, raising water from the Silkstone Seam. A pumping engine, located about one mile south of Sheffield, pumped from the Silkstone Seam at a depth of 40yds, intercepting much of the water, which would have otherwise flowed into the Nunnery Collieries. In addition, a pair of pumping engines were located, side-by-side, in a chamber at the side of the downcast shaft in the Parkgate Seam, for pumping from this level. An identical horizontal pumping engine was also placed at the bottom of the downcast shaft in the Silkstone Seam. The quality of the coal was such that a large washery, capable of cleaning all of the slack, was constructed. At the same time a new heapstead and new screening arrangements were erected.

In the early 1890s a large brick-making works was erected, with a capacity of six million bricks per annum. The clay and fire clay was obtained from the colliery. In order to reduce the charge for land to dump spoil on, the brickworks was adapted to use a proportion of the spoil produced by the colliery.

In the early 1930s electric lighting at the coalface was in use, using power from the mains at 125V. Generally the lighting was used on one side of the face at a time, and when required, transferred rapidly to the other side. The electricity supply was also used to power drills used at the face. Soon after the colliery was taken over by the NCB it was decided to close Nunnery Colliery and concentrate on Handsworth Nunnery Colliery.

Seams Worked:

Seam	**Depth (yds)**	**Thickness (in)**
Parkgate Seam	111	48-54
Silkstone Seam	215	56

Rockingham Colliery

Rockingham colliery was close to the villages of Hoyland Common and Birdwell, south of Barnsley, and eight miles north of Sheffield. The colliery was adjacent to the Manchester Sheffield & Lincolnshire Railway from Sheffield to Barnsley. The seams worked dipped in a north-easterly direction, at a gradient of 1 in 25, and the colliery take was split by a number of north-west to south-east faults

Rockingham was sunk by the Thorncliffe partners, George and Thomas Chambers. The Chambers brothers had purchased the lease for 800 acres of coal from Earl Fitzwilliam in December 1872. Fitzwilliam had decided to dispose of a large area of minerals in the north-western extremity of his estate leasing the coal at £275 per acre for the Silkstone Seam and £175 per acre for the Parkgate. The lease was offered to the Thorncliffe partners, rather than to the Wharncliffe Silkstone Coal Co., as had been expected, as they were already mining adjacent to the leased area and had made a bid for the coal.

Sinking, of a 14ft-diameter upcast shaft and a 16ft-diameter downcast shaft, commenced on 10 June 1873 and progressed well, reaching the Parkgate Seam in March 1875. The 16ft-diameter shaft was named the No.1 or South shaft, and the other was called the No.2 or North shaft. No tubbing was used in either of the shafts.

On 31 March a shot fired during sinking released methane gas, which ignited and set fire to the seam. The fire was followed soon after by a violent explosion. That evening the shafts were sealed to extinguish the fire, but later that evening a further explosion blew off the shaft seal, damaging the No.1 headgear and enginehouse. After flooding of the shafts to extinguish the fires, it was eight weeks before sinking began again and the Silkstone Seam was reached on 14 September 1875 at 339yds. Production started later that same year. The Silkstone Seam was to be a mainstay of the colliery, producing coal until it was exhausted and abandoned in 1946.

The Thorncliffe partners became a limited company known as Newton, Chambers & Co. Ltd, and named their new colliery Rockingham, in honour of the Marquis of Rockingham, a forebear of the Fitzwilliams. At a Directors Meeting in December 1883 the decision was taken to erect a gas and benzol works, comprising 170 beehive ovens located on the north side of the railway. By the mid-1870s over 1,500 tons of coke were produced each week, and experiments with benzol disinfectants were being carried out. Although some valuable pioneering work was carried out by the works, due to the vagaries of the market the works were not economically successful.

From 1885-1886 the mineral lease was increased by a further 2,150 acres. By the 1890s the colliery was working the Silkstone, Parkgate and Thorncliffe Thin Seams. At this time all of the coal was wound from the Silkstone Seam level, the Thorncliffe Seam won by a drift rising from the Silkstone Seam at a gradient of 1 in 5, and the Parkgate Seam by a further drift rising from the Thorncliffe Seam also at a gradient of 1 in 5. Coal from the Parkgate Seam was 'dropped' down a 9ft diameter staple shaft to the Thorncliffe Seam, from where it was transported to the Silkstone level. In the late 1890s the normal daily output from the three seams was; Parkgate 505tons, Thorncliffe Thin 256tons and the Silkstone 553tons. The Barnsley Seam was never worked at Rockingham, having already been extracted from the area leased, before the shafts were sunk.

In 1908 the old beehive ovens were replaced by a battery of forty-five Koppers regenerative ovens built on the old gas works site. An aerial ropeway, one of the first in the country, was used to transport coal from the colliery to the ovens. These ovens worked until 1929.

In late 1908, development of the Top Fenton Seam was begun. During late 1909 an electrically-powered Gillott, rotary disc type, coal-cutting machine was introduced into the Top Fenton Seam. This coal-cutting machine was followed by a second in late 1910.

By 1914 the working faces at Rockingham were approaching the village of Harley. In order to reduce travelling time for men and materials, it was decide to sink a service shaft at Skiers Springs.

Rockingham Colliery, 1979.

Sinking commenced on 1 June 1915 and a shaft was sunk to the Silkstone Seam at a cost of £46,000. This shaft had to pass through the flooded workings of the old Lidgett Colliery and it was necessary to drain the workings by lowering a sinking pump down the old Lidgett pumping shaft. To further secure the Skiers Springs shaft, the Francois Cementation Process was used to consolidate the ground in the region of the old workings, making them impervious, and thereby preventing future accumulations of water from flooding the new shaft. The Skiers Springs shaft also acted as a service shaft for Thorncliffe and Barley Hall Collieries. An electric winder installed at the shaft was the first electric winder in the area, and in 1929 a ventilating fan was installed at the site.

In 1924 a drift was sunk from the surface to the Haigh Moor Seam, the entrance to the drift some 40yds east of the No.2 shaft. The drift did not connect with any underground roadway and served no purpose, until 1942 when workings in the Lidgett Seam were experiencing methane problems, and a further drift was put down from the initial drift to the Lidgett Seam.

The first pulverized fuel-burning boiler, of the modern type, was installed at the colliery in November 1925. Though experimental, the facility was full-size, and was used continuously for the commercial production of steam for use at the colliery.

On 1 April 1925 Newton, Chambers & Co. Ltd purchased nearby Hoyland Silkstone Colliery, which they closed in 1928, working the colliery's reserves from Rockingham. The 508yd deep Hoyland Silkstone shaft became a service shaft, from the Top Fenton Seam level, for Rockingham.

Following closure of Rockingham Colliery's coking ovens, a new overhead ropeway was built to link Rockingham with the new central Smithywood coking plant at Ecclesfield. The ropeway was four and a half miles in length. It was first started on 18 March 1930. During its thirty-five year life, it carried many thousands of tons of coal, before being dismantled in 1965.

During early 1930 the pitch pine headgear at the No.1 shaft was replaced by a steel structure. This was followed by replacement of the No.2 headgear in May 1931. The headgear at the No.2 shaft was manufactured by Newton & Chambers' Thorncliffe Steelworks.

In 1933 a new washery plant was erected, and by 1937 the latest coal-cutting machines and conveyor belts were in use. Pit ponies were still in use at the colliery, but numbers slowly diminished over the years, until July 1956, when the last four ponies were transferred to work at nearby Barrow Colliery.

In 1938 the twin cylinder vertical steam winding-engine, at the No.1 shaft, was replaced. In May 1938 the company introduced a scheme for training young boys prior to employment underground. The scheme provided a six month training course at the company's underground school. The Coalminers' Apprentice Course was one of the first of its kind in the country.

The period from 1932 to 1939 was one in which Rockingham Colliery showed working losses, though the group of collieries owned and run by Newton, Chambers & Co. made a profit as a whole. Investigation into Rockingham's financial position showed that the principal reason for losses was that the colliery had been designed and equipped for a considerably larger output than was being achieved. The question of extending the royalty around the colliery was considered, but in view of the company's preoccupation with the war, and opposition from adjoining colliery owners, it was decided to defer any proposals.

In 1943 a briquetting plant, using coal dust from the screening plant, was established at the colliery. The briquettes were manufactured from high-quality coal and were much in demand. In May 1943 drifts were driven through the Elsecar Fault to the Lidgett Seam, and working of the Lidgett commenced in August 1943. Working of the area of the Lidgett Seam to the dip of the fault proceeded, and continued until December 1952.

Development and operation of the Newton, Chambers Collieries was becoming increasingly complex, and in 1944 it was decided to divorce the collieries from the remainder of the Newton, Chambers business. On 1 July 1945 N.C.Thorncliffe Collieries Ltd, a one hundred per cent subsidiary of Newton, Chambers & Co. Ltd was formed. Reorganization was immediately put into action and the central offices of the colliery undertaking were moved from Thorncliffe to Skiers Springs.

Company estimates of the recoverable reserves of coal remaining at 31 December 1946 amounted to 87,741,100 tons, which at current rates of output would give a life of seventy-five years for the undertaking. A scheme to close Thorncliffe colliery and work the reserves from Rockingham was draw up, thereby effecting economies at Thorncliffe and increasing output at Rockingham. The scheme was known as the 'R. & T. Scheme'. On 2 April 1946 approval to implement the scheme was given by the Ministry of Fuel and Power. Coal mined at Thorncliffe was drawn at Rockingham, that from workings below the Tankersley Fault was hauled up the Harley and Parkgate Drifts, a total distance of 2,200yds uphill. At Vesting Day, on 1 January 1947, the R. & T. Scheme was underway. A new drift linked the Fenton haulage Road at Thorncliffe and the Jubilee Level at Rockingham. Pit-bottom improvements were carried out at Rockingham as well as structural alterations to tipplers and screens at the surface.

In July 1951 it was decided to work the Flockton Seam, lying at a depth of 170yds, from the No.2 shaft at Rockingham. It was decided to work the area of coal, of approximately 440yds by 500yds, using hand-filled retreating longwall units.

In September 1951 an inset was made into the Swallow Wood Seam in the upcast shaft at Skiers Springs. A roadway was driven into the seam and encountered old workings. Examination of available plans showed that coal had been worked in the vicinity of the upcast shaft, from an adit. In addition there were ironstone mines, over 150 years old, which had been worked in the area. After passing through this disturbed area unworked coal was found. A surface drift, at a gradient of 1 in 4, was driven to the Swallow Wood Seam. Roadways and a face were developed and production commenced in late 1952. Coalfaces between the Lidgett and Elsecar Faults were worked between 1952 and 1961. Swallow Wood coal was worked to the dip side of the Elsecar Fault from drifts driven through the fault in 1956 and 1960. Extraction of the Swallow Wood Seam commenced in 1957 and ceased in May 1971. After almost twenty-five years of production the last strip of coal was produced at Skiers Springs on Friday 14 January 1977.

The developments in the Swallow Wood Seam, at Skiers Springs, and the developments in the Flockton Seam, had a marked impact on the output and productivity of Rockingham Colliery. In the week ending 19 December 1953 the colliery produced a record output of 17,000tons per week, and at the year end, an output of 659,909tons, more than double the pre-nationalization figure.

In 1958 a shearer machine was installed at Skiers Springs 228s Face. It was the first mechanized unit to be worked at the colliery. In 1953 pithead baths were installed at the colliery. In December 1956 the first panzer face conveyor and shearer machine was installed in the Top Fenton Seam in the Athersley district of the mine. This proved to be very successful and was adopted in the rest of the colliery as new faces were opened up.

The Harley Seam, which occured in the Skiers Springs shaft at a depth of 9yds, was first worked in April 1956 from two drifts, an intake and a return, driven from the main surface drift. The seam was worked until April 1972 when conditions deteriorated and it was decided to abandon the workings.

The first haulage man-rider was introduced at Rockingham in 1959, comprising four carriages with a twenty-four man capacity, travelling a distance of 1,050yds, soon to be extended to 1,830yds. This was known as the Rockley Manrider. A ski-lift type man-rider was used for a period of time, to carry men up and down a 1 in 4 drift between the Fenton and Parkgate Seams.

The Lilleshall twin cylinder vertical steam winding-engine, which worked at the No.2 shaft, was no longer capable of the pressures placed upon it by constant winding, and it was decided to replace it. During the summer holiday of 1960 it was replaced with an electric winder. The new winder was situated in a new enginehouse built at the rear of the old stone enginehouse.

Three ton mine-cars, replacing the 9cwt tubs in use at the No.2 Pit, were introduced during the summer holiday of 1961. The reorganization had been carried out in order to cope with output from the Low Fenton, Parkgate and Fenton Seams. In addition, underground storage bunkers were constructed in order to smooth the flow of output to the surface.

In August 1964 a fully reserved training face was developed in the Wharncliffe Low Fenton Seam, to train all new faceworkers in the No.5 Barnsley Area. The elusive 1M ton output eventually came on 19 December 1964, when a total output of 1,003,242tons was recorded. Following the reorganization of the No.2 shaft it was then the turn of the No.1 shaft. In October 1965 the 9cwt tubs were replaced by three ton mine-cars.

In 1967 Wharncliffe Silkstone Colliery was merged with Rockingham. The surface plant at Wharncliffe Silkstone was closed and all output was conveyed underground and up the No.2 shaft at Rockingham.

Up to 1970 the output of Rockingham had been relatively stable, but in the early 1970s poor geological conditions were responsible for poor production output, leading to fears for the future of the colliery. Following the introduction of modern mining techniques, such as powered face supports, retreat mining, Dosco heading machines and a new electric winder at the No.1 shaft, the output of the colliery increased and by 1974 was in the region of 700,000tons per annum.

The steam winding-engine at the No.1 shaft was replaced by an electric winder during the summer of 1972, thus replacing the last large steam winding-engine in the Barnsley Area. The engine itself was reported to be capable and efficient, but its boiler plant was said to be worn out.

In mid-1973 the *Colliery Guardian* reported that the Lidgett and Top Fenton Seams at the colliery, which had produced 700,000tons the previous year, were to be abandoned to prevent subsidence damage to a new housing development. The repercussions of this decision were to have a grave impact on the life of the colliery.

At a review meeting on Thursday 3 March 1977, held at Area Headquarters, the Area Director reported that the remaining workable reserves at the colliery amounted to some 1M ton, which would provide the colliery with about two years further life. The Fenton Seam had been abandoned due to the release of massive quantities of methane gas, and it was reported that the Whinmoor Seam was only 17in. thick and not regarded as a workable proposition. There were only small areas of the Swallow Wood Seam remaining to be worked, and these were at shallow depth beneath built up areas. The remaining areas of the Thorncliffe Seam were also of thin section, and where it was being worked beneath Hoyland and Elsecar villages, was causing severe subsidence damage. The Flockton Seam had already been abandoned due to a thick band of dirt in the seam, which caused dust problems and dangerous roof problems.

A Press Statement, dated 15 March 1977, read:

> *The forthcoming closure of Rockingham Colliery, Hoyland Common was announced today by Mr John Keirs, Director of the NCB's Barnsley Area. The coal currently being worked will probably be exhausted before the end of the year and there are no further reserves which can be safely or viably exploited, he said. Jobs for all the 1,170 men who wish to transfer will be available at other collieries in the Barnsley Area.*

The Area Director went on to review the coal remaining in the Rockingham area:

> *Hopes for extending the life of the colliery had been pinned on the Low Fenton Seam', said Mr Keirs. 'But development in the seam had yielded massive quantities of methane gas in the area, which could not be controlled. The conditions were such that the existing workings in the Low Fenton were to be sealed off with an explosion-proof stopping within the next few weeks. There was about one year's coal in the Thorncliffe Seam, but to work it would involve subsidence damage on a massive scale in an area of Elsecar and Hoyland which had twice suffered severe disturbance. The whole of the Thorncliffe coal was under built-up areas and apart from the cost of subsidence damage repairs the cost in human misery was too great.*

Answering trade union questions about coal reserves in the Harley and Whinmoor Seams, the Area Director said that a thorough investigation of the Harley Seam had shown it to be a 'totally unworkable proposition'. In places where the Harley had been worked the men themselves had asked to have the faces closed because of bad conditions. The seam was very close to the surface and could cause severe damage to buildings. In places, the 28in thick seam included 18in of dirt, and it would be impossible to operate faces within the dust regulations. Similar dust conditions would apply in the Whinmoor Seam, but in any case this seam was thin and dirty and was overlain by waterlogged old workings in the Silkstone Seam, only 50yds above. The only entry to this seam would involve the hazards of passing through the waterlogged workings. There were, said Mr Keirs, projects in the Barnsley Area which would offer the Rockingham men far better working conditions and job security than anything left at the 102 year old pit.

There was strong opposition to the proposed closure from the local NUM. Later that year, in July 1977, poor geological conditions seriously affected production in the Thorncliffe Seam. In an effort to save the colliery an exploratory boring programme was initiated, but the results were disappointing, and the fate of the colliery was sealed.

In late 1977 it was common knowledge that the NCB were planning to develop a drift mine at the Hemsworth Colliery site at Fitzwilliam, to be known as the Kinsley Drift Mine. Negotiation between the local NUM and the Area Director secured an agreement for the Rockingham workforce, who wished to do so, to be transferred to the new mine en bloc. As the colliery worked out its remaining reserves and production ran down, men were gradually transferred to Kinsley Drift. Between June 1978 and July 1979 about fifty men were transfered, and on Friday 10 August 1979 a further fifty men were transferred, as the first face at Kinsley came on stream to produce coal.

On Wednesday 20 June 1979 V17s face sheared its last strip of coal from the Thorncliffe Seam and, later that afternoon, the last mine-car of coal from this seam was wound up the No.1 shaft. There was then only one piece of coal to work and that was in the Low Fenton

Seam on the S56s face. This could not be worked until powered supports were salvaged from V17s for use on S56s. By 27 June S56s was producing coal. S56s continued to produce coal into November. On Monday 19 November a fault was reported on the tailgate of the face and, on the following Tuesday, the last strip of coal was sheared. This was wound the following day, up the No.2 shaft, and despatched for power station consumption. This was an historic day for the colliery, which had produced over 50m tons of coal in its 104 year life.

On Friday 23 November 1979 a further seventy-five men were transfered to Kinsley Drift, and most of the few remaining men were employed on underground salvage at the colliery.

The working relationships between management and men at Rockingham had always been good. This was well recognized in the locality, and was demonstrated by the number of long-serving employees. One employee, Tom Wood, a shaftsman and brickie at the colliery, had worked for the company for over sixty-two years.

Seams Worked:

Swallow Wood Seam, 3ft thick, 25yds deep, extensively worked at Rockingham and Skiers Springs since the early 1940s

Lidgett Seam, 3ft 1in thick, 97yds deep, only small areas worked at Rockingham and Skiers Springs.

Flockton Thick Seam, 5ft thick, 169yds deep

Top Fenton Seam, 3ft 2in thick, 225yds deep

Low Fenton Seam, 2ft 11in thick, 229yds deep, developed in 1934 but soon discontinued due to inferior quality, reworked during the Second World War.

Parkgate Seam, 4ft 6in-6ft thick, 249yds deep, furnishing a steam-coal. Extensively worked providing a good quality, thick section, coal-seam.

Thorncliffe Thin Seam (often refered to as the Swilley Seam), 2ft 10in thick, 282yds deep, furnishing good quality house and gas-coal.

Silkstone Seam, 5ft thick, 339yds deep, furnishing best house and gas-coal.

Stocksbridge Colliery

Stocksbridge Colliery was developed by the steel-making company of Samuel Fox at Stocksbridge, to the north-west of Sheffield. Steel-making began in the early 1860s, the new industry generating a demand for coal. The Stocksbridge valley was well situated for the supply of local coal, gannister and fireclay, which were worked predominantly by drift mines, because of the steep sided nature of the valley.

Fox's first mine was one which had been opened by predecessors of theirs, which they worked until the early 1860s. Fox's first colliery, known as the North No.1 Pit, was a drift mine. In about 1910 a further drift was driven about half a mile to the west of the North No.1 Pit, to supplement the output from this pit. This later drift became known as the New Pit, and the No.1 Pit was referred to as the Old Pit. These drifts were sunk to tap the Halifax Soft Seam, the lowest of the commercially workable seams, which provided a first-class coking and gas-coal.

Mining of the Halifax Soft Bed started at Stocksbridge Colliery in 1861. From 1870 a small area of the Halifax Hard Bed was also worked until its abandonment in 1916. In 1916 a new pit was started, the old workings continuing for a few years more until they were abandoned in 1928.

In 1937 it was decided to drive a drift into the workings. Work proceeded for a period of time but was stopped by the problem of water ingress and flooding. The colliery had always experienced problems with water.

When the colliery was taken over on Vesting Day, 1 January 1947, it was recognized to be in poor condition. However, the large reserves of both Halifax Hard and Soft coal in the Don Valley, combined with a large mining-based workforce, were felt to be encouraging. Reserves totalling 11.4m tons were estimated, of which in excess of 6m tons were in the Halifax Hard Seam. The plan was to enter the Halifax Hard Seam and work this coal, to support new developments driven into the Soft Bed, in order to replace current workings which were exhausted. As the scheme proceeded it became apparent that machine-mining was not feasible, and that without machine-mining it would not be economically viable. Interspersed in the seam were large numbers of ironstone boulders, up to 3ft in length. The boulders meant that it would have been impossible to machine cut the coal, due to the risk of sparks igniting gas and the danger of injury to the cutter man.

In the prevailing circumstances the newly formed NCB had little choice but to close the colliery. An NCB press release, dated 22 April 1948, stated, 'the Board has no alternative but to close the pit and the men have been informed of the reasons through the consultative machinery.' It was decided that there would be no redundancy, and that the 268 men employed would be transferred to neighbouring pits. The majority of the men employed at the pit lived locally, mostly in Stocksbridge and Deepcar.

Seams Worked:

Halifax Hard Bed Seam, outcropping, 30in thick.
Halifax Soft Bed Seam, 55yds deep, 24in thick.

Thorncliffe Colliery

Located five and a half miles north of Sheffield, Thorncliffe Colliery was located within the substantial area occupied by the Thorncliffe Ironworks of Messrs Newton, Chambers & Co., on a constricted site adjacent to the LNER. Newton, Chambers & Co. owned the colliery until nationalization.

The Parkgate Seam was the first to be worked on a large scale at the colliery. A sloping drift, started in 1859, was driven at a gradient of 1 in 10 to the seam. The drift had been driven to connect with workings from earlier pits in the vicinity and by the mid-1890s had reached a length of 1,100yds.

A geared steam engine, located at the mouth of the drift, drove an endless-rope haulage system on the main incline, while general haulage underground was performed by heavy carthorses.

A battery of 153 beehive coke ovens was situated close to the colliery. One hundred of these ovens were patent ovens, which produced by-products, as well as generating steam and providing coke. The by-products included oil, ammonia water and gas. The oil was treated in a distillery to produce; light burning oil, light naphtha, light cylinder lubricating oils, heavy lubricating oils for large bearings, paraffin wax for candles, pitch and, last and most important of all, Izal. In 1885 an analytical chemist called J.H.Worrall came to Thorncliffe to undertake research into the analysis of gas from beehive coke ovens. During this work, his attention was turned to the liquid product of the gas and then the oil produced. From the distillate he produced an oil with powerful germicidal properties, which was registered under the trade name Izal in 1893. Used to manufacture soaps of various qualities, Izal was claimed to be the most effective disinfectant known at the time.

A bank of boilers, fired by waste heat from the coke ovens, provided steam for the haulage engine, surface workshops and the ammonia works.

In 1924 the New Silkstone Drift was driven. The drift enabled a large area of the Fenton Seam to be worked. Thorncliffe Colliery was connected to Skiers Springs Pit by means of the Fenton Seam. The Barley Hall shaft served as a second outlet and, as an upcast ventilation shaft and for pumping. By the mid-1930s the Parkgate, Fenton and Silkstone Seams were being extensively worked at the colliery, with coalface machinery in use in the Fenton and Silkstone Seams.

In the 1940s there were two drifts at the site, the Parkgate Drift and the New Silkstone Drift. Because of the congested nature of the site, a pair of aerial ropeways were used to carry spoil away from the screening plant.

On 24 September 1943 the *Yorkshire Post* reported:

> ***YORKSHIRE MINERS' FINE RECORD OF OUTPUT***
> ***17 Men Raise 47,700 Tons of Coal in Nine Months***
>
> *The 47,700 tons of coal had been raised between the 9th November 1942 and 16th August 1943, by the 17 men who had maintained an average of 12.58 tons per man-shift, working in the Fenton Seam at the colliery...*

The achievement had been recorded in a panel in an exhibition opened by Lloyd George, the Minister of Fuel & Power, which was to begin a tour of Yorkshire collieries the following week. The panel noted that 37,850tons of coal was sufficient to equip an army division with tanks, guns, ammunition and lorries, that 7,000tons would coal the ships required to move the division to a destination in Europe and that 2,850tons would help to make the bombers and fighters that would provide air cover to the convoy.

In 1945 the Company were planning to close the colliery and concentrate output at the Barley Hall Colliery site. Finally, on 2 April 1955, after some 100 years of operation, the last coal was drawn at Thorncliffe Drift.

Seams Worked:

Top Fenton Seam

Low Fenton Seam

Parkgate Seam, averaging 5ft thick, including a 3in dirt band, the seam yielded 4ft 9in of workable coal. The 1ft thick Brazils section of the seam yielded a first rate house-coal which was popular in the London market, where it found a ready sale. Formerly, this section of the seam was frequently not removed. One of the old check-weigh men is reported to have said 'Oh, sir, I should think there are hundreds and thousands of tons buried; it's a great shame, and one a't best kitchen coals i't world.' The 2ft 4in thick Hards was turned into a high grade coke, which was principally used in the ironworks, and a small quantity sold. The remaining 1ft 5in thick Softs produced a gas-making and an engine coal, Sheffield providing the principal market for these coals.

Thorncliffe Seam

Top Silkstone Seam

Whinmoor Seam

Thorpe Hesley Colliery

In 1900-03 Newton, Chambers & Co. sank the No.1 Pit, an 18ft-diameter pumping shaft, on the north-eastern edge of the village of Thorpe Hesley. Thorpe Pit was sunk to drain the Norfolk and Smithywood Colliery workings and as a means of access for the men. Newton, Chambers & Co. were to own the colliery until nationalization.

On 31 December 1899 a pair of horizontal tandem compound condensing differential pumping engines were ordered from Messrs Hathorn Davey & Co. of Leeds, order number 5712. The two engines were handed (one left handed, one right handed), but otherwise identical. The two large pumping engines were delivered in July 1902, and installed at the surface to operate pumps at the bottom of the shaft, by means of timber pump rods. At the time of installation the pumping facility was reputed to be one of the finest in the country.

In 1914 the No.2 shaft, a 14ft diameter shaft, was sunk at Thorpe Hesley to provide an upcast for Smithywood Colliery. This shaft was sunk below the Silkstone Seam and connected by a level drift to Silkstone workings on the dip side of the Tankersley Fault.

Thorpe Hesley downcast shaft, 1974.

The section through Thorpe Pit:

Seam	Depth (yards)	Abandonment date
Top Fenton	38	November 1964
Parkgate	67	November 1962
Thorncliffe	86	June 1967
Top Silkstone	167	October 1968
Low Silkstone	170	June 1928
Jack	190	December 1964
Whinmoor	223	January 1967

At a later date the pit became a service pit for Smithywood Colliery, for man-riding and supplies. No coal, except for a small quantity used in the boilers, was drawn at the colliery. A statement of pumping, dated 1935, showed the average quantity pumped in summer months to be 706,000 gallons per day, and the maximum capacity of the pumping plant to be 5,364,000 gallons per day.

Following closure of Smithywood Colliery, in December 1972, Thorpe Pit reverted to a permanent mine pumping station, with submersible pumps in the shaft replacing the differential pumps.

Wentworth Silkstone Colliery

Wentworth Silkstone was a drift mine located three and a half miles south-west of Barnsley. The colliery was owned and worked by Wentworth Silkstone Collieries Ltd, and was known locally as Levitt Hagg or 'T Levvy.

Two drifts were sunk around 1911-12, at a gradient of 1 in 5, almost certainly to work the Silkstone Seam, from which the colliery derives its name. By the late 1940s the colliery was working the Flockton and Fenton Seams. At the time of nationalization the workings were some one and a half miles in-bye, and the men had to walk to and from their places of work. At Vesting Day the colliery was producing an output of 500tons per day, which gradually increased to 900-950tons by mid-1949. On Wednesday 10 August 1949 the colliery produced 1,000tons per day for the first time. Pithead baths were opened in June 1949.

Output had previously been raised to the surface by means of three separate haulages. A scheme to convey all output via a common underground trunk-conveyor system was installed, and a new coal preparation plant erected. Coal was conveyed from the coalface to the screens via some 3,000yds of trunk-conveyor.The work was completed by the late 1950s having cost some £500,000. As a result of the reconstruction, annual output rose from 159,000tons to 344,00tons.

Water was frequently a problem, and the mine regularly had to deal with some twelve tons of water for every ton of coal produced.

In the period 1955-1959 the two drifts were extended to work the Whinmoor Seam. In the mid-1960s the colliery was producing an average of 8,000tons per week from the Whinmoor Seam and was regarded as a class 'A' colliery, likely to remain in production for the foreseeable future. With the shift toward thinner seams, an area in which the colliery had extensive experience, it was felt that the colliery had a future. At this time the colliery was

Wentworth Silkstone Drift, April 1980.

working seams from 2ft 3in to 2ft 10in in thickness. It was selling its output to the CEGB, after washing and blending with other coals which were brought in. An article in the Sheffield Star, dated 11 November 1961, reported:

> ***ANOTHER PIT IS MECHANIZED***
> *Another Barnsley area colliery, Wentworth Silkstone, where 650 men are employed, will be fully mechanized by the end of this year. When the scheme is complete there will be two mechanized faces in the Whinmoor Seam, producing power station coal, and one in the Fenton Seam, from which comes coal for coking and household purposes.*

Later in the 1960s there were fears that the colliery would close and the *Sheffield Star* reported on 4 December 1967:

> ***NCB DENY PIT CLOSURE RUMOUR***
> *Men at the Wentworth Silkstone drift mine, which employs about 600, have been worried by rumours predicting an early closure because of poor grade coal.*
> *But a spokesman at the NCB headquarters in Doncaster said: No proposals for the closure of this mine are currently being considered by the regional board*

The colliery eventually ceased coal production and closed in June 1978.

Seams Worked:
- Flockton Seam
- Fenton Seam
- Whinmoor Seam, 2ft 3in to 2ft 10in thick.

Wharncliffe Silkstone Colliery

Wharncliffe Silkstone Colliery was situated on the western edge of the South Yorkshire coalfield, in a quiet valley near the hamlet of Pilley, some four miles south of Barnsley, on the Sheffield & Barnsley Railway.

The colliery royalty comprised an area of approximately 2,000 acres, the majority of the seams to be worked occurring in the middle coal measures, the Whinmoor Seam alone in the lower coal measures.

The Great Northern Railway Co. had acquired the land around Wharncliffe Silkstone, which lay close to their line from Sheffield to Barnsley, as one of their colliery properties. However the Railway Co. Act did not allow them to engage in the coal trade, and colliery properties owned by railway companies were taken over by private individuals, usually connected with the railways. In the case of Wharncliffe Silkstone this was Mr Robert Baxter, who was solicitor to the Great Northern Railway Co. and Mr George Blake Walker of Sheffield. These were later joined by Mr Edmund Baxter, Mr Horace Walker and John and Parkin Jeffcock (of Messrs Woodhouse & Jeffcock, Mining Engineers of Derby). Parkin Jeffcock was later to lose his life in the Oaks Colliery Disaster of 1866 as one of the rescue party. He is remembered in a memorial, close to the site of the disaster, in Barnsley.

So the Wharncliffe Silkstone Colliery was born, taking its name jointly from Lord Wharncliffe, who was the chief royalty owner, and the Silkstone Seam, which was the first seam to be worked. The Wharncliffe Silkstone Colliery Co. were to own and work the colliery until nationalization.

Sinking commenced in 1853 under the direction of Messrs Woodhouse & Jeffcock. The winding pits were situated to the south-east of the royalty. Sinking proceded without difficulty and all of the shafts were brick lined throughout, without the need for shaft tubbing.

The No.1 shaft was sunk to the Silkstone Seam. It was 151yds in depth and 12ft in diameter. This shaft became the outlet for coal from the Silkstone, Top and Bottom Fenton, Parkgate, Thorncliffe Thin and Flockton Seams. The winding-engine at the No.1 shaft was manufactured by Davis of Tipton in 1855. This engine was a single cylinder, 33in diameter by 5ft stroke, vertical winding-engine, winding using flat ropes on 12ft diameter reels. The engine was counterbalanced by a bunch of chains, connected to the main drum-shaft by a flat wire rope wound onto a small counterbalance drum. The counterbalance rose and fell in a 60ft deep staple shaft at the rear of the engine house. This arrangement of winding-engine was common in the coalfields of Northumberland and Durham but unusual for South Yorkshire.

The No.2 shaft was about half a mile away to the south-west and was sunk 196yds to the Whinmoor Seam. This shaft served as an upcast for part of the Whinmoor workings. The main upcast, or No.3, shaft was sunk to the Thorncliffe Thin Seam at a depth of 90yds and was 15ft in diameter. The No.4 shaft was originally sunk to the Silkstone Seam, at 165yds. In about 1890 it was deepened to the Whinmoor Seam, at a depth of 220yds. The diameter of the shaft was also increased to 13ft 6in when the shaft was deepened. This shaft was the outlet for the Silkstone and Whinmoor Seams and was the main pumping shaft for the colliery. When originally sunk, a beam pumping engine, with a 65in diameter cylinder and a 78in stroke, raised water from the sump at a depth of 165yds in three bucket lifts.

About one and a half miles from the main No.1 shaft was the Hermit Hill shaft. This 12ft-diameter shaft was sunk to the Whinmoor Seam at 98yds, to provide access to an area of the

Wharncliffe Silkstone, early 1900s.

seam and reduce travelling time. Some three quarters of a mile north of the No.1 shaft a day drift was located at Athersley. Shaft Nos 1, 3 and 4 were located 30-40yds apart. Close to shafts No.1 and No.3 was a pumping shaft, which was sunk to the Silkstone Seam at a depth of 151yds. Silkstone coal was raised in No.1 shaft, Thorncliffe Thin coal in No.3 shaft and Whinmoor coal in No.4 shaft from early in 1897.

The faulted nature of the coalfield fragmented the area worked by the colliery into a large number of separate districts connected by stone drifts cut through the faults.

The colliery opened in 1854, in a period of depression and the early years were notable for trade disputes. In 1858 the colliery owners in the district gave notice of a fifteen per cent reduction in wages. As a result of a secret meeting of local men, led by a John Normansell of Wharncliffe Silkstone, the South Yorkshire Miners' Association was formed on 10 April 1858. John Normansell was appointed as the first checkweighman in the history of the coal industry, following the Royal Commission of 1868. Wharncliffe Silkstone Colliery was proud of the fact that it was the No.1, and oldest, branch of the National Union of Mineworkers, Yorkshire Area.

During the early days of the colliery George Blake Walker, the son of Horace Walker, was studying mining engineering under Professor Attwell. Walker visited Germany and gained a wealth of knowledge of Germany mining technology and methods. In 1871 he returned to England where he worked at a large colliery concern in Durham, adding further to his mining knowledge. Later he came to Wharncliffe Silkstone and undertook the duty of undermanager of the Thin Seam, where he devoted a lot of energy to remodelling the underground workings and installing the latest equipment. George Blake Walker was a pioneer, and through his efforts the company became pioneers in mining technology and methods and led the South Yorkshire mining industry. Walker was elected a director of the Company in 1882 and Chairman in 1888.

Some of the important innovations employed at the colliery include:

1882: The first compressed air coal-cutter.
1890: Use of electrically driven coal-cutters.
Introduction of miners' hip baths.
1892: Introduction of the first large coal washing plants in South Yorkshire. This plant was on the Luhrig system and greatly improved both the quality and output of the beehive type coke ovens in use at the time. On 12 May 1962 the *Barnsley Chronicle* reported:

> ***Oldest washery in Yorkshire coalfield***
> *Wharncliffe Silkstone's 109 years-old colliery has the oldest working washery in the Yorkshire coalfield – and they are proud of it.*
> *Built by the Germans, its Luhrig washer has washed 40 tons of coal an hour for over 70 years without a major breakdown*

1897: The 160 old-type, beehive coke ovens were replaced by by-product ovens, erected in conjunction with Messrs Simon Carves Ltd. The 'waste' gas from these ovens was used to power gas engines at the colliery. The use of waste gas to power gas engines became extensive at the colliery, the engines used to produce electricity. The coke oven plant erected by Messrs Simon Carves Ltd was the latest of its kind in the country, and started production in 1899. Comprising a battery of thirty-five coking ovens and by-product recovery plant, the plant produced coke, tar, benzol and sulphate of ammonia.
1899: Breathing apparatus was introduced, long before mining legislation made their introduction mandatory.
1902: Instigation and staffing of the first joint rescue station, erected at Tankersley.
1907: The first coal conveying plant was installed.

By the late 1890s seventeen Gillott & Copley coal-cutters were employed to undercut the Thorncliffe Thin, Silkstone and Whinmoor Seams, all of which were worked on the longwall principle. The machines were powered by compressed air, supplied from air compressors on the surface. From 1912 electric power was generated at the colliery by gas engine driven generators and a mixed pressure turbine. The gas engines were powered by coke oven gas from the company's adjoining coking plant.

Underground haulage was by endless rope, powered by a steam engine located on the surface. A well-appointed workshop provided for the repair of underground and surface machinery and plant, and a blacksmiths shop carried out general maintenance and repairs.

The majority of the coal raised at Wharncliffe Silkstone Colliery produced excellent coke as well as being rich in by-products. Soon after the colliery was sunk a battery of beehive type coke ovens was erected. These produced an excellent grade of coke and, despite the waste of by-products, a number of them were retained to manufacture a special quality of metallurgical coke.

Developments and experience in coal carbonization proceeded rapidly and in 1912 fourteen more ovens were erected. These were regenerative type ovens, utilising the waste gas in gas engines, which powered generators producing electricity for consumption at the colliery. This was the first time that power had been produced in this way in Yorkshire. Subsequently ten more regenerative ovens were built.

The Miner's Welfare always held an important place in the consideration of the colliery management. The Institute in the colliery village promoted concerts and had facilities such as billiard tables. There was also a library connected with the Institute. Under the terms of the 1920 Mining Industry Act a fund, known as the 'Betterment Fund', was created by allocating one penny per ton of colliery output, for the purpose of welfare.

A brickworks connected to the colliery initially manufactured bricks for use in and about the colliery. The brickworks was later upgraded and began to manufacture first class building bricks, roof tiles and ridges, which found a ready market in Yorkshire and the South of England.

A landslide on the spoil heap at the colliery in January 1952, which took nine months to consolidate, was followed by a further landslide, in late 1952, and subsidence on the roadway at Pilley from underground workings, which closed the road.

By the early 1950s there were only three shafts in use at the colliery, and these were used solely for ventilation and pumping. At this time the Fenton Seam was being worked, and coal from the seam horizon was transported to the screening plant via a short surface drift. Declining coal reserves and deterioration of the quality of the coal had compelled the mining engineers at the colliery to look at the large areas of high-quality coal that lay 'sterilized' under built-up areas. By 1957 the only reserves left in the Flockton Seam were either inferior quality or pillars left to support surface property. It was decided to work these pillars, and pneumatically stow the goaf with dirt, as machine mining proceeded. During 1954 a 1 in 1 gradient surface drift had been driven to the Flockton Seam, and this was used to facilitate stowing in the seam. The scheme was successful and a great many lessons were learned for the future.

Coal winding at one of the oldest collieries in Yorkshire ceased on 28 January 1966, and the remaining coal was drawn at Rockingham Colliery. The winding machinery at Wharncliffe Silkstone Colliery was in need of extensive repair and it was more economical to draw the coal at Rockingham Colliery.

Seams Worked:

Flockton Seam, 2ft 5in thick, 9yds deep.

Top Fenton Seam, 3ft 5in worked, principally furnished a gas-coal and also made good coke, 32yds deep.

Low Fenton Seam, 4ft 2in worked, 36 yds deep.

Parkgate Seam, 5ft 3in worked, 60yds deep, furnished good coking coal, gas and steam-coal as well as being rich in by-products.

Thorncliffe Thin Seam, 4ft 6in worked, 90yds deep, provided a valuable house gas and steam-coal, the small-coal made a good coke.

Silkstone Seam, 4ft 9in worked, 156yds deep,

Whinmoor Seam, 221yds deep, 2ft 9in section, mainly provided a gas-coal and also good coke

7
Mining 1851-1875

By 1870 the momentum in the coal industry of South Yorkshire was beginning to increase. Output had risen to 4.4m tons from 108 collieries, though at the time West Yorkshire was still ahead, with an output of 6.75m tons from 308 collieries. The statistics show that the average South Yorkshire colliery was larger and produced a significantly higher output than those in West Yorkshire.

Though there were a large number of small enterprises still working in South Yorkshire, many of the large-output collieries of the twentieth century were sunk in this period. These include collieries such as Barrow, Denaby Main, Manvers Main, Houghton Main and Woolley. Denaby Main Colliery, at over 440yds to the Barnsley Seam, was the deepest and most westerly colliery in Yorkshire when it was sunk on the edge of the concealed coalfield.

The typical size of royalties worked at this period varied from 1,500 to 3,000 acres, with the collieries further to the east working larger royalties. In the case of the sinking at Denaby Main, this amounted to a royalty of 7,000 acres.

The development at this phase shows the relentless eastwards development of the coalfield, following the valleys of the rivers Dearne and Don, and the sinking of deeper, more capital-intensive, collieries. It also shows a gradual opening-up of the central region of the coalfield between Barnsley and Doncaster. Many collieries at this time were sunk to work the Barnsley Bed Seam at depths between 200 and 500yds, usually 300yds, in the Middle Coal Measures.

The Frickley and Maltby Troughs were still undeveloped and collieries were yet to be sunk in the concealed region of the coalfield.

1851-1875 was a period of great advance for South Yorkshire, establishing their pre-eminence in the coal mining industry. During this time many of the large collieries, which were to become the backbone of the coalfield until nationalization and beyond, were sunk. Many were to work for a hundred years and more, before they were closed because of exhaustion of their reserves.

As well as the deep shaft mines, drift mines, such as Dearne Valley, were sunk, at a later date, specifically to work the shallow Shafton Seam.

Collieries sunk in this phase include:

Aldwarke Main
Cortonwood
Darfield Main
Dearne Valley (drift mine)
Denaby Main
Houghton Main
Kilnhurst (also known as Thrybergh Hall)
Kiveton Park
Manvers Main
Mitchell Main
Monk Bretton

Orgreave
Treeton
Waleswood
Wath Main
Wharncliffe Woodmoor 1, 2 and 3
Wharncliffe Woodmoor 4 and 5
Wombwell Main
Woolley

Aldwarke Main Colliery

Aldwarke Main Colliery was situated two miles north east of Rotherham, about one mile from Parkgate, near the Rotherham to Rawmarsh road. The colliery was initially leased by Messrs Waring & Co., who leased the rights to the Barnsley Bed Seam. It was connected to, and served by, the LMS and LNER.

The Barnsley Seam at the colliery, was almost level and of good quality. Two shafts, located about 40yds apart, were sunk to the Barnsley Seam at 150yds. The seam furnished 8ft of clean coal, and was in production by 1865.

In 1873 Sir John Brown & Co. of Sheffield, acquired Aldwarke Main and Carr House collieries, and in the period 1890-1893 sank Rotherham Main. Aldwarke Main becoming the principal colliery owned by the company, remaining in their ownership until nationalization. John Brown immediately set about developing the lower seams at Aldwarke, and by 1877 had sunk to the Swallow Wood Seam at 217yds, and the Parkgate Seam at 405yds.

The company's colliery concerns were an important part of their operation, as shown in 1899 when the company employed 3,500 at their Atlas Works and 4,500 at their collieries.

There were four shafts sunk in two pairs at Aldwarke, which divide the colliery into two mines – the Parkgate Mine and the Swallow Wood-Barnsley Mine. At the Parkgate Mine the No.1, downcast and No.2, upcast, shafts were sunk to the Parkgate Seam at 405yds, and some time later the No.1 shaft was further deepened to reach the Silkstone Seam at a depth of 494yds. Both shafts were 18ft in diameter and lined with tapered brick, except for a short length (6yds) of cast iron tubbing inserted where a thin seam of coal and water occurred. In 1879, 1884 and 1890 exploratory work was carried out from the pit-bottom of the No.1 shaft, which had been deepened to the Silkstone Seam.

The No.1, Parkgate Pit, wound Parkgate Seam coal, and the No.2, Parkgate Upcast Pit was fitted as a drawing shaft, but normally served as a ventilating and man-riding shaft for the Parkgate Seam.

At the Swallow Wood/Barnsley Mine the No.3, downcast, and No.4, upcast, shafts were sunk to the Swallow Wood Seam, at a depth of 217yds. Both shafts were 14ft in diameter. The No.3, or Swallow Wood Pit, wound Swallow Wood Seam coal, and the No.4, or Swallow Wood/ Barnsley Upcast Pit, served as the ventilating pit for the Swallow Wood and Barnsley Seams, as well as for man-riding and raising Barnsley Seam coal. Coke was manufactured at the colliery by 220 beehive type coke ovens, which also provided waste heat to a bank of Lancashire boilers.

On 14 June 1883 members of the Gas Institute were entertained at the colliery. A luncheon was served on the main haulage level in the Parkgate Seam, to the strains of a full orchestra.

Between 1893 and 1894 a shaft was sunk at Warren House, some two and a half miles north-west of Aldwarke Colliery. The shaft was sunk to the Parkgate Seam and used for ventilation and man-riding. Between 1908 and 1909 a second shaft was sunk at the Warren House site. This shaft was sunk to the Swallow Wood Seam and was also used for ventilation and man-riding.

Between 1906 and 1912 experiments were carried out on two coal briquette plants, but the trials came to the conclusion that the plants were not financially viable.

In 1907 the colliery was electrified. In the same year coal was mechanically washed at the colliery. In 1924 the Silkstone Seam was reached by a stone drift from the Parkgate Seam, and Silkstone coal was wound up the No.1, downcast, Parkgate Pit. In the same year the pit bank was remodelled and an up-to-date screening plant erected.

Some years before it was decided to extend the life of the Parkgate Seam at the colliery by developing a large area of the seam, lying to the south of the shafts beyond the Southern Don Fault. The Don Fault runs from the middle of Sheffield to beyond Doncaster, approximately parallel to the River Don. Between the shafts and the Northerly Don Fault the strata were practically flat, but to the south of the fault they dipped steeply. This, together with a series of small faults, had the effect of throwing the Parkgate Seam down to a depth of some 900yds, compared with 404yds at the shafts. To win the Parkgate Seam in the new area it was decided to drive drifts at a regular gradient, to reach the coal at a point where the seams were again level, rather than follow the seam. Two drifts were driven at a gradient of 1 in 3 in the late 1930s, and a large endless-rope haulage system was installed underground, to work one of the drifts.

Up until 1931 the colliery had produced some 35m tons of output from the three seams in production. The Parkgate royalty at the colliery extended three and a half miles from the pit-bottom and, to reduce the travelling time, the men were conveyed in specially designed 'mail corves' for some two and a quarter miles of this distance.

In the early 1940s the Swallow Wood, Parkgate and Silkstone Seams were being worked, the Silkstone by a drift from the Parkgate Seam. All coal was wound up the No.1 downcast shaft from the Parkgate level, at a depth of 405yds. The No.2 upcast shaft was 494yds deep to the Silkstone level. At the time the working coalfaces were a long way from the shaft.

In June 1960 ninety redundancies were announced following the proposed closure of the Parkgate Seam. The majority of the men made redundant were re-deployed to the New Stubbin and Manvers Main Collieries. On 30 June 1961 the colliery was closed due to exhaustion of its seams.

Seams Worked:

Barnsley Bed Seam, 150yds deep, 7ft 6in thick, producing a good quality steam-coal.

Swallow Wood Seam, 216yds deep, averaged 5ft in thickness, principally used for gas manufacture, though it also furnished a good quality steam-coal.

Lidgett Seam, 255yds deep, 2ft 6in thick.

Parkgate Seam, 405yds deep, varied from 5ft to 5ft 6in in thickness of 'clean' coal without a parting and furnished a first rate gas-coal.

Silkstone Seam, 494yds deep, 2ft 8in thick, some faults in the seam.

Cortonwood Colliery

Cortonwood Colliery was situated a short distance from Lund Hill Colliery, (the scene of a mining disaster in 1857 which killed 189 miners), five miles from Barnsley, and five and a half miles from Rotherham.

The Cortonwood Colliery Co. was formed in 1872, under the name 'The Brampton Colliery Co.', specifically to exploit the Barnsley Seam on the Elsecar branch of the Dearne & Dove Canal. The deed of partnership, dated 12 July 1872, was drawn up in Lancashire, between: Charles Bartholomew of Doncaster; Samuel Roberts of Sheffield; Robert Baxter of London; Henry Davis Pochin of Surrey; Benjamin Whitworth of London; Thomas Whitworth of Manchester; John Devonshire Ellis of Sheffield; James Holden of Manchester and William Pochin of Miles Platting. Each of the partners was required to contribute £20,000 into the concern immediately, and up to £30,000 maximum, as required.

Sinking of two of the largest shafts in South Yorkshire commenced in October 1873, by the owners Cortonwood Collieries Ltd, who owned the colliery until nationalization. The two shafts were sunk to a depth of 238yds, reaching the Barnsley Seam at 213yds in March 1875. During sinking, tubbing was inserted in each shaft to a depth of 85yds, in order to prevent the ingress of water from old workings in the Meltonfield Seam, which lay close to the surface. The No.1 Whitworth downcast shaft was 20ft diameter, and the No.2 upcast shaft was 15ft diameter. The tubbing in the downcast shaft was believed to be the largest in England at the time it was installed. Barnsley Seam coal was raised in the No.1 downcast shaft.

In 1876 'quite a village of concrete cottages' was erected near the junction of the Elsecar Canal Branch, at a place called New Wombwell. 'It may be stated that the whole of the houses, 116 in number, are built of concrete even to the roofs and spouts.' These cottages survived until some time around 1960.

In October 1883 a Memorandum of Association of the Cortonwood Colliery Co. Ltd was drawn up, with a total capital of £175,000 made up of £10 shares.

Cortonwood Colliery, April 1984.

Between 1907 and 1908 both shafts were deepened to 506yds in order to work the Parkgate Seam at 481yds. At the same time the company decided to remodel the surface plant and install the latest coal preparation plant. In 1908 the company dismantled their beehive ovens and placed an order for a battery of fifty of the latest, patent-regenerative type ovens from Koppers Coke Oven & By-Products Co. of Sheffield. They also ordered coal compressing machines and by-product recovery plant and extensive new sidings were laid out. The total cost was about £50,000. The new plant produced coke, tar, sulphate of ammonia, benzol and surplus gas. It was reputed to be unequalled in performance by any other plant in South Yorkshire.

The method adopted for sinking the pits from the Barnsley to the Parkgate Seam was somewhat novel, as the work was carried out continuously without interference to normal coal winding operations. The upcast shaft was first deepened to the Parkgate Seam. A heading was then driven from the upcast shaft to the underside of the downcast shaft, 26yds below the Barnsley Seam. A connection was made upwards and a substantial scaffold erected beneath the sump in the downcast shaft. A ventilation bypass was installed in the heading and a winding-engine at the surface operated a sinking bucket in the shaft. Sinking debris in the downcast shaft was brought out along the heading and up the upcast shaft, so that it did not interfere with coal winding in the downcast shaft. The Parkgate Seam was well laid out for dealing with a large output, with straight main roads and gradients which favoured loaded trams.

In the period 1925-27 the No.1 Whitworth downcast shaft was deepened to 606.5yds, to intersect the Silkstone Seam, which was developed in 1927 and which, with the exhaustion of the Parkgate in 1941 and Swallow Wood in 1975, became the mainstay of the colliery. In the late 1930s the deepening of the No.2 upcast shaft was begun, reaching its final depth of 602.5yds by 1940.

From 1873, until its exhaustion in 1921, the Barnsley Seam provided the bulk of the colliery's output and was wound up the No.1 shaft. From 1908, until its exhaustion in 1941, Parkgate coal was wound in the No.2 shaft. In 1918 the Haigh Moor Seam was developed and initially wound in the No.2 shaft, but with the exhaustion of the Barnsley Seam in 1921, it was then wound in the No.1 shaft. The Haigh Moor Seam was worked until its exhaustion in 1975.

A major reconstruction scheme was carried out over an eight-year period from 1927 to 1935. The main features of the scheme were concentration of coal winding by skips in the No.1 shaft, and installation of electric winding-engines, compressors and coal preparation plant. In addition, the underground transport system was reorganized using trunk conveyors.

In 1927 development of the Silkstone Seam began. Coal from the seam was transported via a drift to the Parkgate level, from where it was wound in the No.2 shaft. Part of the 1927-35 reconstruction scheme was the installation of a clutch-drum electric winder in the No.1 shaft, enabling Silkstone coal to be raised from the Parkgate level. With the exhaustion of the Parkgate Seam in 1941, Silkstone coal was again wound in the No.2 shaft, though from the Silkstone inset.

In the early 1940s, both shafts were coal winding. Haigh Moor coal wound from the Haigh Moor level in the No.1 downcast shaft, and Silkstone coal from the Silkstone level in the No.2 upcast shaft. In 1958 grouting work was carried out in the No.1 Whitworth downcast shaft to consolidate the shaft lining.

In 1961 Silkstone coal was once again wound in the No.1 shaft, together with Haigh Moor coal. Coal from the two seams was wound in batches. In the 1960s a modern coal preparation plant was constructed at the colliery by Head Wrightson Minerals Engineering Ltd. In total some £1million had been spent on the new coal preparation plant, electrification and skip-winding at the colliery.

In the 1970s the colliery was working both the Swallow Wood and Silkstone Seams, but by the early 1980s the colliery was working the Silkstone Seam alone. It was producing coking coal, which could not be sold due to a slump in the market for coke for the steel industry.

The proposed closure of the colliery in 1984 marked the start of the year-long national miner's strike. Following the return to work in March 1985, the colliery remained idle and was finally closed in October 1985, with a manpower of 802: 633 underground workers and 169 surface workers. On 26 October 1985, the *Daily Telegraph* reported:

> ***CLOSED PIT THAT STARTED MINERS' STRIKE***
> *Cortonwood Colliery, the South Yorkshire pit whose threatened closure sparked the miners' strike, produced its last coal yesterday after 112 years. During the 11 month dispute the pit's ramshackle picket hut, known as the Alamo, became a symbol of defiance and 'no' surrender... The NCB announced the intended closure of the pit early last year after losses over four years had totalled £12 million. The pit was losing £20 for every ton of coal produced...Voluntary redundancy has been taken by 320 of the 690 men at the pit and 370 have transferred to nearby long-life pits at Maltby, Barnburgh, Silverwood and Treeton.'*

Seams Worked:

Barnsley Seam, 213yds deep, worked from 1873 this seam was finally worked out in 1921.

Haigh Moor Seam, (or Swallow Wood), 270yds deep, this seam was developed in 1918 and exhausted in 1975, a medium coking coal with a high proportion of high ash coal called 'Jabez', 3ft 8in of the 4ft 4in seam thickness was worked.

Parkgate Seam, 481yds deep, development began in 1908 until exhaustion in 1941.

Silkstone Seam, 571yds deep, developed in 1927, a strongly coking coal, 2ft 8in thick, of which the full seam section was worked. Working in the 1970s.

Darfield Main Colliery

Located at Low Valley, midway between Wombwell and Darfield village, four miles south-east of Barnsley. Due to its location the colliery was known locally as 'the Valley'. The colliery sidings had direct access to the main line running from Manchester to Doncaster. The seams at the colliery generally dipped to the north-east at a gradient of 1 in 13.

Sinking of two shafts commenced in 1856 and the 7ft 6in. thick Barnsley Bed was reached in July 1860. The shafts, an 11ft 6in diameter No.1 downcast, and a 10ft 4in diameter No.2 upcast shaft, were sunk to a depth of 337yds. Both were lined with cast-iron tubbing to a depth of 163ft, no doubt as a protection against the water bearing strata known locally as the Meltonfield Rock. Mining of the Barnsley Seam commenced soon after sinking was complete and continued until its exhaustion in 1910.

In October 1872 a serious underground fire resulted in work at the colliery being suspended for twelve months or more, with a loss to the owners in excess of £100,000.

Prior to nationalization the colliery was owned by a number of colliery owners, including; the Darfield Coal Co., Shipley Collieries (who owned the colliery twice during this period) and the Mitchell Main Colliery Co., who owned the colliery immediately prior to nationalization.

The neighbouring Mitchell Main Colliery, two thirds of a mile to the west, was a rival to Darfield until the two were amalgamated in the early 1900s, under the ownership of the Mitchell Main Colliery Co. This amalgamation increased the gross 'take' of the two collieries to 3,000 acres, a block of minerals extending three and a half miles from east to west and one and a half miles from north to south. The Mitchell Main shafts had been deepened to the Parkgate Seam in 1899 and the decision was taken to work seams below the Barnsley Seam from Mitchell Main and those above from Darfield Main Colliery.

In 1913 a decision was taken to sink a new shaft below the Barnsley Seam to the Thorncliffe horizon level. A continental firm won the contract and sinking proceeded until halted by the First World War. In 1917 sinking resumed and the shaft proceeded to completion. The 21ft diameter No.3 shaft was completed, reaching the Thorncliffe Seam at 625yds, with insets into the Meltonfield, Winter, Beamshaw, Swallow Wood, Lidgett, Flockton, Fenton and Parkgate Seams. The shaft was notable for its use of what was known as 'German' tubbing for the first 300ft of sinking, reputed to be one of the best examples in Yorkshire. The new shaft was used to augment ventilation at Mitchell Main Colliery, and to improve coal winding and ventilation at Darfield Main, becoming the upcast for the pit, the existing No.1 and No.2 shafts reverting to downcast shafts.

Exhaustion of the Barnsley Seam led to development and working of the Meltonfield Seam from 1910 onwards, and the Beamshaw from the late 1930s. A report, dated 19 February 1931, on the viability of Mitchell Main and Darfield Main Collieries by Hugh F. Smithson, came to the conclusion:

> *I have grave doubts in my mind as to the desirability of maintaining Darfield Main at all, as a coal winding proposition... the output... does not warrant extensive replacements by modern machinery... The economy resultant from drawing the whole output of the two collieries at Mitchell shafts would be considerable.*

The plan was to formulate a scheme to effect the abandonment of Darfield Main, except for pumping and ventilation purposes, and possibly for the winding of men employed in the more easterly districts. In the event it was Mitchell Main that closed some twenty-five years later!

In 1943 the No.2 shaft ceased to be used for winding and served as an upcast for ventilation. The headgear at the shaft was dismantled in 1949, in the modernization scheme that was carried out in the period 1948–54. No.3 shaft became the main coal winding shaft for the colliery, as well as serving as the upcast ventilating shaft for part of the old Mitchell Main workings and Darfield workings in the Meltonfield, Winter and Beamshaw Seams.

In the early 1950s a further reappraisal of the two collieries was carried out and the decision was taken to close Mitchell Main. Following this decision, reorganization was carried out at Darfield. The surface reorganization involved the construction and electrification of the winding-engines and houses, and construction of coal preparation plant, pithead baths and railway sidings. Underground, additional faces were developed, principally in the newly exploited Winter Seam, to employ manpower transferred from Mitchell Main. The objective was to achieve a weekly output of 10,000 saleable tons per week at an output per man-shift of 30cwt. By 1959 the modernization and reconstruction were complete and the targets achieved. The colliery, which was producing good quality coking coal, was 'secure for 100 years to come'.

Darfield Main Colliery, 1980.

Mitchell Main closed in 1956, reverting to pumping and became known as Mitchell Pumping Shafts. Mitchell Main manpower was transferred to Darfield, boosting the workforce to about 1,300, on the Meltonfield, Winter and Beamshaw Seams. All coal at Darfield was won from conventional longwall faces, machine cutting and hand filling the coal.

On 22 November 1956, soon after a dispute at the colliery, the *Sheffield Telegraph* reported an all-time output record of 12,000tons in a week. The weekly target at the time was set at 10,000tons. Self-advancing supports and trepanner coal-cutters were introduced into the Meltonfield Seam in 1961, resulting in a revolution in coal winning methods. By 1963 all hand-filled faces had been phased out and replaced by fully mechanized units. Geologically the Winter Seam was the least conducive to mechanization, being thin and having an overburden of soft bind, which made the support of the roof difficult with mechanical mining.

The Meltonfield, Winter and Beamshaw Seams were in close proximity to each other, which caused problems for future developments as it was a contravention of the 1954 Mines and Quarries Act to have two or more seams working in the same locality at the same time. In addition, large areas of Meltonfield goaf were waterlogged and had to be drained before the Winter and Beamshaw Seams below could be worked. In the western part of the colliery take, the close proximity of the water-bearing Meltonfield Rock resulted in that area of the colliery being particularly wet. To the east of the take, the bind overburden was thicker, resulting in the seam being drier and even dusty in some areas.

The colliery was split into two parts, a northern and a southern area, by the Darfield Fault, which traversed the take from east to west. The Lundhill Fault, to the east, formed the eastern boundary with the adjoining Wath and Cortonwood Collieries. To the north, the Edderthorpe Fault formed the boundary with the adjoining Houghton Main Colliery.

In 1970 extraction in the Meltonfield Seam ceased, the Beamshaw and Kent Thick Seams providing the output at the colliery. Retreating faces were first used at the colliery in 1975.

From the 1970s the quality of the coal from the Beamshaw and Kent Thick Seams deteriorated. This resulted in the traditional markets for coking coal being closed to Darfield. The colliery was then tied to the requirements of a sole customer, the CEGB.

By December 1978 the whole of the colliery output was produced from retreat mining. By the early 1980s a development was underway to exploit the Silkstone Seam, at a depth of 695yds, by deepening the No.2 and No.3 shafts and integrating the colliery into the 'South Side' mining complex centred on Grimethorpe Colliery, some four miles to the north-west. The reconstruction scheme cost £23million, and was complete by the mid-1980s. In the early 1980s the Kents Thick Seam was abandoned and production concentrated in the Silkstone Seam. In November 1986 Darfield was merged with Houghton Main, and finally closed in July 1989.

Seams Worked:

Seam	Depth (yards)	Thickness (inches)
Meltonfield Seam	170	48-54
Winter (or Abdy) Seam	196	30
Beamshaw Seam	220	36
Kents Thick Seam	270	42
Barnsley Seam	337	90
Fenton Seam	575	54
Parkgate Seam	600	60
Silkstone Seam	695	36

Dearne Valley Colliery

Dearne Valley Colliery was situated five miles east of Barnsley, and ten miles west of Doncaster. Located near the village of Great Houghton, the colliery was a few hundred yards from the adjoining Houghton Main Colliery. The Dearne Valley Colliery Co. was formed in 1901 with a capital of £30,000.

Sunk in 1901, the colliery worked the Shafton Seam, which furnished coal especially suitable for railway steam-raising and found a ready market with British Railways. Two drifts were sunk in 1900, followed by a shaft in 1901, which was used for ventilation. At Dearne Valley the Shafton Seam was disturbed by a fault with a 60yd downthrow, and to access the seam, cross measure drifts, at a gradient of 1 in 8, were driven across the fault.

Dearne Valley Colliery was very much a 'family' pit, many of the men having worked there all their lives, and the labour relations had always been exceptionally good. For many years the colliery had one of the highest output rates in the coalfield, at about 30cwt per man-shift. This was despite the fact that the colliery was one of the last to be mechanized, the coal being manually trammed to the gate ends and ponies were still being used underground. In the early 1940s the overall output per man-shift at the colliery was 27.02cwt, against a county average of 24.51cwt. A pumping station on the colliery premises supplied water to the Dearne Valley Water Board.

The Shafton Seam was typically 4ft 8in thick, standing on a dirt parting 2in thick, beneath which was a band of 4-6in of an inferior coal, which was sometimes worked to provide boiler fuel. In the early 1940s the full seam thickness was being worked. A particular difficulty with Shafton Seam coal was its abnormal friability, which made stocking impracticable. Stocking trials were carried out unsuccessfully in the early 1930s, but were not repeated.

During the 1950s and 1960s the changeover from steam to diesel locomotives resulted in a slump in demand for locomotive coal. The slump in demand resulted in large stocks of coal building up, and it was decided to reduce the manpower by transferring men to the adjacent Houghton Main Colliery. Dearne Valley employed some 600 men and the plan was to transfer sixty men, and possibly as many as 100 men ultimately. On 5 January 1960 the *Yorkshire Post* reported:

SURPRISE MOVE FOR 100 DEARNE MINERS
Pit is hit by BR switch to Diesels
About 100 men at Dearne Valley Colliery, Little Houghton, near Barnsley, have learned that that they are to be transferred to neighbouring Houghton Main Colliery, although Dearne Valley was not on the original list of pits where the labour force was to be cut... Demand for the pit's special type of coal has diminished as a result of the railways' increased use of diesel oil and big stocks are piling up in the pit yard...

It had been planned to close the colliery when the last face finished in the Shafton Seam in September 1972. However, the news in the *Colliery Guardian* read 'Dearne Valley Reprieved'. The colliery was to be given an extra fifteen years life, and money found to open up the Sharleston Top Seam and two further seams in the future. Two drifts had already been driven and a new face was due to start as the last face in the Shafton Seam closed. In 1973 a further drift was sunk for men and materials. By the mid-1980s the colliery was working the Sharlston Yard Seam, producing coal for the power station market.

In 1982 a new ventilation shaft was sunk to work additional reserves at the colliery. The ventilation shaft was part of a scheme to improve the infrastructure of the roadways, haulage and transport systems, and cost £4million. The scheme was completed by the mid-1980s, and enabled Dearne Valley coal to be transported down a borehole and connect with the Grimethorpe coal clearance system, for winding and processing at Grimethorpe.

Dearne Valley Colliery worked on as part of Houghton Main, and from the mid-1980s as part of the 'South Side' complex centred on Grimethorpe Colliery. By the early 1990s prospects were not looking good for the colliery. It was working the thinnest seams in the area, and in the words of its manager, 'has to run very hard to stand still'. In the late 1980s the colliery was working the Sharlston Yard Seam, and had a production target of 350,000tons in 1990, from a labour force of 255 men. On 1 March 1991 the *Daily Telegraph* reported:

700 to be laid off as pits close
More than 700 jobs are to be scrapped with the closure of two pits, British Coal announced yesterday... Losses of £2million were recorded at Dearne Valley since April last year, a figure expected to rise to £3.5million by the end of the financial year. Mr Siddall (Group Director) said he could see no way the pit could produce coal profitably for a sustained period.

Dearne Valley Colliery closed on 5 March 1991, and by 1992 the site was cleared.

Seams Worked:
Shafton Seam, 5ft 4in
Sharlston Yard Seam
Sharlston Top Seam

Denaby Main Colliery

Denaby Main was situated six miles south-east of Doncaster, on the south bank of the River Don Valley. The mineral property under lease extended from Swinton to Doncaster, a distance of seven miles, covering an area of almost 7,000 acres. The colliery occupied a constricted site, sandwiched between the road, river and railway line. It was served by the LNER, and Denaby Staithes on the South Yorkshire Navigation Canal.

The sinking of the colliery was a high-risk pioneering venture by the established West Yorkshire coal owners, Messrs Pope & Pearson. When sunk, the colliery was the deepest and most easterly in Yorkshire. The workings of the colliery were beneath the magnesian limestone and it had previously been the opinion of many mining engineers that the coalfield terminated where the line of the limestone began. No other Yorkshire coal owners had taken the same magnitude of risk in sinking a colliery as did Pope & Pearson, and this was not to be equalled until the sinkings in the Doncaster area some forty years later. The Denaby Main Colliery Co. was registered in March 1868, with a share capital of £110,400. Five men owned almost all of the shares.

The two shafts at Denaby Main were about 40yds apart. The No.1 downcast shaft was 14ft in diameter, and the No.2 upcast shaft 13ft 6in diameter. In order to stem the heavy feeders of water, which were encountered close to the surface, the two shafts were lined with cast iron tubbing, from the surface to a depth of 70yds, and below that, with brick down to the 9ft thick Barnsley Seam, at 447yds. The shafts were located between two major faults known as the Northern and Southern Don Faults, which were approximately parallel, running in a north-east to south-west direction. The two faults were approximately 3,000yds apart and both were downthrown to the south-east. The Northern Fault had a throw of 120ft, and the Southern Fault 380ft.

Coke, sold principally to crucible steel works, was manufactured by seventy-two beehive type coke ovens at the colliery. A range of workshops provided facilities to repair and manufacture wagons and machinery. A granary, with six silos, provided storage for the food required for over 400 horses, which would eventually be employed between Denaby Main and Cadeby Main Collieries. In 1893 a new company, the Denaby & Cadeby Main Collieries Co. Ltd was formed. In March 1927, the company was amalgamated into the Yorkshire Amalgamated Collieries Ltd. On 27 April 1936 the company was incorporated under the title the Amalgamated Denaby Collieries Co. Ltd, a title which was retained until formation of the NCB on the 1st January 1947.

The colliery was adjacent to the Manchester, Sheffield & Lincolnshire Railway (which became the Great Central Railway in 1897) from which coal was despatched. In the late nineteenth century, following a long-standing feud between the colliery and railway company over alleged poor service, the colliery was looking for alternative transport options. At the end of the nineteenth century, the colliery company built two staithes on the navigable river Don, which ran adjacent to the colliery. Early in the twentieth century, the two staithes were merged into one large staithe. A regular customer for Denaby coal, transported by the river Don, was Rotherham power station. A further staithe was added in 1954, which survived until the 1980s.

Working of the Barnsley Seam was limited to the coal bounded by the two Don Faults. The first coal from the Barnsley Seam was raised in 1865 and the seam was worked manually by pick and shovel working, into tubs, until abandonment in May 1968. By that time some 3,600 acres of the seam had been worked, yielding 30m tons of coal. The seam was up to 9ft 7in in thickness, and 6ft 6in to 7ft 3in of the section was worked. The Barnsley Seam, as at other collieries in the district, was liable to spontaneous combustion.

Output increased rapidly, and in 1897 the colliery raised an output of 629,947tons in 281 working days, a record tonnage of coal drawn at a single shaft from any British colliery. The largest output had been 2,673tons in one day, a record that the owners hoped to increase to 3,000tons. In 1912 the No.2 upcast shaft was deepened from the Barnsley Seam level to the Silkstone Seam at 798yds, and at the same time a 14ft diameter, 360yd deep, staple shaft was sunk from the Barnsley level to the Parkgate Seam. The Silkstone Seam was never worked at Denaby and the shaft was filled to the Parkgate level at 707yds. The Parkgate Seam varied from 4-5ft in thickness, and was worked over approximately the same area as the Barnsley Seam, though not as extensively. Parkgate coal was wound in the No.2 upcast shaft and Barnsley coal in the No.1 downcast shaft.

The Parkgate Seam was worked from 1912 until 1964, yielding 13m tons of coal. The Parkgate Seam varied from 4ft 6in to 5ft in thickness, and the full seam section was worked. From 1956 Parkgate coal was transported, on a small scale, along the underground connecting road between Denaby and Cadeby Collieries, to be wound at Cadeby. By July 1958 half of Denaby's output was being wound at Cadeby, and by 1959 Denaby ceased coal winding.

As at the Barnsley Seam, hand won methods gradually gave way to semi-mechanized forms of mining. Working faces 200yd-long were undercut, blown down with explosives and hand filled onto conveyors which loaded the coal into tubs. The tubs were then hauled by endless-rope to the shaft bottom. Later, these were replaced by mine-cars hauled by flameproof diesel locomotives.

Between 1920 and 1931 the surface plant at the colliery was extensively re-modernized. Up until the mid-1920s output from the Parkgate Seam was wound up the staple shaft to the

Denaby Main Colliery, after merger with Cadeby in 1968.

Barnsley level, where it was taken to the No.1 shaft for winding to the surface. The No.2 shaft had previously been used for ventilation and safety access only. The expense of sinking a coal-drawing shaft to the Parkgate Seam would have been formidable, and the decision was taken to install simultaneous decking in the small-diameter No.2 upcast shaft, to enable Parkgate coal to be drawn directly up the shaft to the surface. In about 1931 a new steam winding-engine and concrete headgear were installed at the No.2 upcast shaft. The No.1 downcast shaft was then used solely for drawing Barnsley Seam coal.

In the 1940s a subsidiary gas works at the colliery supplied gas to local housing in Denaby village, as well as to the local munitions works.

By the early 1940s problems were being experienced, with frequent roof falls at the coalface when water was present in the overlying clod, and poor geological conditions in the Parkgate Seam were giving rise to uncertainty about the future of the seam. Because of this uncertainty the Haigh Moor Seam was being developed. An inset into the Haigh Moor Seam was made in the staple shaft at a depth of 497yds and coal from the seam was lowered to the Parkgate level, where it was wound to the surface. The Haigh Moor Seam was 4ft 3in thick and the full seam section was worked. From the 1940s, output at the colliery declined due to low output in the Parkgate Seam, caused by the poor conditions. Until nationalization there were no pithead baths at the colliery.

Following nationalization a reconstruction scheme was initiated to concentrate output from Denaby and Cadeby Collieries at Cadeby No.2 shaft, using a clutch-drum electric winder to allow skip-winding from two horizons. By 1956 Cadeby was winding all of the coal produced by Denaby Main. The cost of the reconstruction was estimated at £4.3million, and was completed in early 1959.

In May 1968 Cadeby and Denaby Collieries were amalgamated, and Denaby Main Colliery ceased to exist as a separate unit. The Barnsley Seam had been worked continuously during the life of Denaby Main Colliery. The seam had been worked with pick and shovel and hand got into tubs, no conveyor belt, coal-cutting machine or shot-firing was used in the coal getting process.

When the colliery closed it had produced in excess of 50m tons of coal, costing the lives of 203 men. Following closure, the concrete upcast headgear and winder were retained as an emergency exit for Cadeby Main Colliery.

Seams Worked:

Barnsley Seam, 448yds deep.
Haigh Moor Seam, 497yds deep, 4ft 3in thick.
Parkgate Seam, 707yds deep, 4ft 6in-5ft thick.

Houghton Main Colliery

Situated five miles east of Barnsley, and ten miles west of Doncaster, the area worked by the colliery comprised some 2,500 acres between the villages of Darfield and Great Houghton. Houghton Main was sunk by the Houghton Main Colliery Co., who operated the mine until nationalization. In 1875 the company were registered with a capital of £100,000.

Sinking of two brick-lined shafts, both 14ft diameter, commenced in 1871 and was completed to the Barnsley Seam, at 527yds, in 1873. The No.1 shaft was the downcast and No.2 was the upcast shaft. The colliery lay beside the main LMS Railway line between Sheffield and Leeds, from which it was served by sidings. This route provided a connection with the LNER.

In 1909 a 12ft 6in diameter staple shaft was sunk from the Melton Field Seam, at 347yds, to the Barnsley Seam, at 516yds, to lower output from the Melton Field to the Barnsley Seam, from where it was raised to the surface.

From 1919-21 the No.2 upcast shaft was deepened from the Barnsley Bed to the Parkgate Seam, at a depth of 774yds, and in 1925 a further staple pit was sunk from the Barnsley Bed to the Parkgate Seam. This shaft was 16ft in diameter and 256yds deep. Coal winding from the Parkgate, by way of this staple pit, began in 1925. The seams in the colliery take had an average dip of 1 in 12.

In order to work the Parkgate, and other seams below the Barnsley Seam, it became necessary to sink a third, larger-diameter, shaft. The original shafts were too small in diameter for ample ventilation or to raise sufficient output. In 1924 sinking of the 20ft diameter No.3 shaft was started. Sinking proceeded slowly, due to uncertain market conditions, and stopped in 1928. By June 1931 the shaft had reached a depth of 540yds. Sinking recommenced in 1937 and was finally completed in 1940, reaching the Thorncliffe Seam at 818yds. Trouble with water was anticipated and, to counteract this, the shaft was lined throughout with a monolithic concrete lining. The new headgear, heapstead and airlock were manufactured from reinforced concrete. The new shaft was an upcast, and on completion of sinking, the No.2 upcast shaft was converted to a downcast. The two existing shafts then became downcasts for all the seams.

The No.1 shaft was 527yds deep, reaching to the Barnsley Bed, with a 15yd deep sump. The No.2 shaft had a diameter of 14ft, and was 818yds deep to the Thorncliffe Seam, with an increased diameter of 15ft below the Dunsil Seam, at a depth of 546yds. The No.3 shaft was 818yds deep, and 20ft in diameter, to the Thorncliffe Seam, having a 5yd deep sump.

The 16ft diameter Parkgate staple pit was 259yds deep, from the Barnsley Bed to the Parkgate Seam, and a further 36yds to the Thorncliffe Seam. The 12ft 6in diameter Beamshaw staple pit, between the Melton Field and Beamshaw Seams, a distance of 62yds, and to the Beamshaw and Barnsley Bed, a further 109yds.

Coking plant comprised fifty Otto waste-heat ovens, with a capacity of 330tons of dry coal per day, to which were added ten Simon Carves regenerative ovens in 1928, with a capacity of 100 tons of dry coal per day. Waste heat and surplus gas generated was used in a number of boilers at the colliery.

In 1927 the colliery was connected to the Carlton Main Collieries Power Scheme to enable it to receive electricity from the central power station at Frickley.

The Miner's Welfare Committee erected, and formally opened, a fine range of pithead baths in January 1931. Built at a cost of £25,000 the baths had accommodation for 2,016 men.

Until the early 1930s Parkgate coal was raised to the Barnsley Seam level, up the 16ft-diameter staple pit, and from there, up the No.1 downcast shaft.

In the early 1930s coal was produced from the Melton Field, Beamshaw, Barnsley Bed and Parkgate Seams. The output varying from 3,000 to 4,000tons per day. With a workforce of over 2,000, the colliery worked an area comprising some 2,500 acres. The Barnsley Bed was worked by hand throughout its life, while the Melton Field and Parkgate Seams were machine mined.

Following nationalization it was decided to group Grimethorpe and Houghton Main together, to concentrate output at Grimethorpe Colliery. The two collieries were working in a large rectangular area of high-grade coals. It was decided to reorganize and restructure the

Houghton Main Colliery, July 1992.

collieries so that coal from the Meltonfield, Winter and Beamshaw Seams could be hauled by locomotive to Grimethorpe No.1 shaft, for skip-winding to the surface. A locomotive roadway was to be driven to Houghton Main to collect Houghton Upper Series coals. The No.3 shaft at Grimethorpe was to be deepened to the Parkgate/Silkstone series. Houghton Main No.3 shaft was equipped for skip-winding and was to continue winding Parkgate coals for the near future, until the output from the two collieries was raised at Grimethorpe, leaving Houghton Main No.1 and No.2 shafts for man-riding and supplies. Winding in the Parkgate staple shaft was abandoned and output taken by skip-winding from August 1953.

At Vesting Day, on 1 January 1947, output at the colliery was produced from the Melton Field, Beamshaw and Parkgate Seams, via the three shafts and two staples. The layout was such that each seam was, in effect, a separate small mine. The total output of the combined seams was 2,500tons with a manpower of 2,600. At Vesting Day the company had a capital of £654,021.

Face mechanization was introduced and output increased significantly. Output from the Parkgate Seam almost doubled. In March 1950 it was decided to close down the coking plant at Houghton Main, since the plant, which had been in operation for many years, was neither as efficient or economical as other plants in the area. The closure was part of the long-term policy of the Board, which took account of the needs of the gas industry. Some 130 men were employed at the plant and every effort was taken to absorb as many of them as possible.

In August 1953 ventilation at the colliery was reorganized, when the No.3 shaft became the upcast and the original shafts, Nos 1 and 2, became downcast shafts. In 1960-61 a ferro-concrete Koepe winder was built over the No.2 shaft, and the buildings around the shaft, including the disused Guibal and Waddle fan-houses, were demolished.

In the mid-1970s development work was in progress to open up the Newhill and Thorncliffe Seams. Access to the Newhill Seam had been gained by driving a pair of rising drifts, at a 1 in 5 gradient, from the Melton Field Seam main airways. Access to the Melton Field Seam was by cross measure and in-seam drivages from existing workings in the Beamshaw horizon.

In 1977 a £16million investment was approved to exploit reserves of over 7m tons. The scheme involved the driving of underground drifts into the 50 to 72in thick Dunsil Seam. Dunsil Seam output was wound at Houghton No.1 shaft, the coal preparation plant and rail facilities at the colliery handling output from the Dunsil Seam and Dearne Valley Colliery. Lower seam output was conveyed underground for winding at Grimethorpe.

In the early 1980s a new coal preparation plant was built, followed by the construction of a new fan drift and fan-house at the No.3 shaft in the mid-1980s. From the mid-1980s coal was no longer brought to the surface at Houghton, the colliery becoming part of the Barnsley Area 'South Side' project based on Grimethorpe Colliery.

In November 1986 Houghton and Darfield Main Collieries were combined. The combined mine was capable of an output of 1.5m tons from a workforce of 1,235 men. At Grimethorpe a new drift and a coal preparation plant were constructed in the period 1979-84, to process the combined output from Grimethorpe, Houghton, Darfield Main and Barnsley Main Collieries. In the mid-1980s Houghton Main was working the Dunsil, Fenton and Silkstone Seams, but by the late 1980s the situation was not looking good for the colliery. In June 1990, the *Coal News* reported:

> ***CRITICAL TIMES AT LOSING PITS***
> *...And he* [Mr Siddall] *expressed serious concern about Houghton Main Colliery, which is performing well below its output potential. It has the equipment and the face room to produce coal profitably but operating losses last year were £2.9million – and so far this year are already over £2million,' said Mr Siddall.*

The colliery was working the Fenton and Silkstone Seams at this time, and the planned output for 1990 was 1.1m tons from a labour force of 900 men.

In October 1992 British Coal announced thirty-one closures and the loss of 30,000 jobs, with ten collieries earmarked for immediate closure. One of these ten collieries was Houghton Main. On 30 October 1992 the colliery ceased production. In September 1993 it was one of the collieries offered to the private sector but, unfortunately, did not receive any bids to continue mining. As a result British Coal moved swiftly to close the colliery. In late 1993 Houghton Main closed with the loss of over 440 jobs.

Following closure an attempt was made to preserve the complex of surface buildings. The case made was that the surface remains comprised an almost intact range of buildings, typical of a colliery sunk at the height of the mining boom in the Yorkshire coalfield. The attempt was unsuccessful, and sadly the surface remains were demolished and the site levelled.

Seams Worked:

Newhill Seam, 6ft 6in thick.
Melton Field Seam, 347yds deep, 4ft 2in thick.
Beamshaw Seam, 406yds deep, 3ft thick.
Barnsley Bed Seam, 516yds deep.
Dunsil Seam, 533yds deep, 50-72in thick.
Fenton Seam, 736yds deep.
Parkgate Seam, 783yds deep.
Thorncliffe Seam, 818yds deep.
Silkstone Seam, 939yds deep.

Kilnhurst Colliery

When originally sunk, the colliery was known as Thrybergh Hall, only becoming known as Kilnhurst Colliery from about 1923. It was situated at Kilnhurst, about three miles north-east of Rotherham. Comprising a mineral property of some 7,000 acres, the colliery take covered almost four square miles and was bounded by the Northern and Southern Don Faults. This extensive royalty contained 60m tons of proved workable coal. The coal measures dipped to the east from the Don Fault at a gradient of 1 in 20, which increased to a maximum of 1 in 2.5. Both the Barnsley and Parkgate Seams at the colliery were affected by washouts, the Parkgate suffering to a greater extent. A washout in the Barnsley Seam formed the south-western boundary on the dip side of the shaft.

The colliery was adjacent to Kilnhurst station on the LMS and the LNER, to which the colliery sidings were connected. The South Yorkshire Navigation Canal also ran adjacent to the colliery.

The colliery was originally sunk by J.&J. Charlesworth Ltd of Wakefield, who sank two shafts to the Barnsley Seam in 1858. The two shafts were 12ft 6in diameter to the Barnsley Seam at 275yds. When the No.2 shaft was deepened, sometime before 1931, its diameter was increased to 13ft to the Haigh Moor Seam, at 355yds, and 14ft to the Silkstone Seam, at 653yds. The No.1 shaft was the downcast pit and the No.2 the upcast pit, both shafts were used for coal winding. In addition a 10ft diameter pumping shaft was sunk to the Kents Thick Seam, at a depth of 225yds, in about 1870. This shaft was known as the No.3 pumping pit. A Cornish pumping engine operated on the shaft, raising water from the Kents Thick Seam, from about 1870. The engine worked regularly until the Second World War, and was last run in about 1947.

Barnsley coal was raised at the colliery from about 1860. At the time, the colliery was not connected to a railway and coal was transported by barge on the adjacent Sheffield & South Yorkshire Navigation Canal, or carted for local use. Coal transported by canal principally went to Hull and Grimsby, for shipping.

In 1923, the colliery was taken over by Stewarts & Lloyds Ltd of Glasgow, who used the colliery to supply large quantities of coke to their iron and steel works at Corby, and to their works at Bilston and Islip.

In early 1920 the Haigh Moor Seam was developed and opened out. Between 1920 and March 1934 the seam was worked in the gently sloping part of the take, to the north of the shafts. Between 1939 and March 1944 the seam was worked on the south side of the shafts, in the area of maximum dip. In 1944 the seam was abandoned on the south side of the shafts, due to the extremely complicated haulage arrangements required.

The Parkgate Seam was first developed in October 1927 and was working until at least the late 1940s. The workings lay to the south of the shafts, and were worked in a south-west direction up to the washout. Headings were driven through the washout in order to locate the seam again. The haulage arrangements in the Parkgate Seam initially comprised a main haulage way some 750yd-long called Hart's Drift, which ran at a gradient of 1 in 3.3.

In the early 1930s Messrs Stewarts & Lloyds undertook an extensive reorganization of their business, erecting new steelworks and coke ovens at Corby and making improvements to their colliery at Kilnhurst. Up until the 1930s the surface arrangements of Kilnhurst Colliery were little changed from the original layout. During the early 1930s new screening and washing plant were erected and an aerial ropeway was installed to remove spoil from the pit and waste

Kilnhurst Colliery, November 1988.

from the screens and washery. In addition to the surface improvements, the underground operations were greatly extended.

The Silkstone Seam was first worked in about 1931, with the first workings in the gently sloping coal measures to the north-west of the shafts. Further districts were developed on the north and south sides of the shafts. Those on the south side ventured into a steep area and were stopped so that output could be concentrated in the Denaby area of the Silkstone Seam, which was in an area of gently dipping reserves.

In the mid-1930s feed-water for the boilers and for general purposes was still raised by the original beam pumping engine, known as the Kent's pump, which was manufactured in 1846. The pump raised 350gallons per minute, at six strokes per minute, from the Waterloo and Wathwood Seams.

A thriving brickworks was at work in the early twentieth century, and in the mid-1930s some 150,000 bricks were produced each week, using locally quarried clay.

In the early 1930s both shafts were winding coal. The No.1 downcast, was sunk to the Barnsley Seam at 282yds and wound Barnsley coal and Haigh Moor coal. Coal from the Haigh Moor Seam travelled up 1 in 5 gradient drifts from the seam 76yds below, to be wound at the Barnsley level. The No.2 upcast shaft, sunk to the Silkstone Seam at 660yds, wound Silkstone and Parkgate coal. Workings in the Parkgate and Silkstone Seams were connected to the Barnsley Seam by a cross-measure drift, and all the intake ventilation for the lower seams passed through this drift.

In the early 1930s it was decided to shut down the colliery in order to entirely re-arrange and re-equip the surface and underground facilities at the colliery. The colliery was closed for eight months while this work was undertaken.

In 1935 the Kents Thick Seam was developed to the east of the shafts. The seam was practically virgin in the area, and produced a good house-coal. The seam was worked for a number of years, but difficulty was experienced in passing over the pillars left in the Barnsley Seam below. In March 1945, when the colliery was taken over by Manvers Main Collieries, the seam was closed in order to concentrate production in the Parkgate and Silkstone Seams.

In 1936 the colliery changed hands once again, when it was taken over by the Tinsley Park Colliery Co. of Sheffield. The Tinsley Park Colliery Co. sank the 20ft-diameter No.4 downcast shaft. The shaft was sunk to the Silkstone Seam at 642yds and was not equipped for coal winding. There were insets in the shaft at the Parkgate and Silkstone levels, for ventilation purposes only.

In the early 1940s the colliery's output performance was significantly below the county average, the limiting factors being the faulted royalty, which necessitated the working of five coal-seams at the same time in order to maintain output. At this time the Kents Thick, Barnsley, Haigh Moor, Parkgate and Silkstone Seams were all being worked. Kents Thick, Barnsley and Haigh Moor coals were wound at the Barnsley level in the No.1 shaft. Parkgate and Silkstone coals were wound at the Silkstone level in the No.2 shaft. The weekly target at this time was 8,608tons, and the average daily output in September 1941 was 1,253tons.

Working conditions in late 1941 were difficult, due to the considerable faulting being encountered on one of the Silkstone faces, and the distance of the workings in the Barnsley Seam, which were more than two miles from the shaft bottom, and were accessed by a tortuous haulage road.

In the late 1940s the reserves at Kilnhurst Colliery were estimated at 46.75m tons, at which time the Parkgate and Silkstone Seams were being worked. Manvers Main Colliery Co. were in negotiation with the United Steel Co. for the acquisition of Kilnhurst. The United Steel Co. had acquired the Tinsley Park Colliery Co. in December 1944, and took over the management of the colliery from 1 February 1945. The obligation for the supply of Kilnhurst coking coal to Stewarts & Lloyds, from then on, rested jointly with the Manvers Main Colliery Co. and the United Steel Co. Kilnhurst's royalty of 2,750 acres was then added to that of Manvers Main Colliery. In the late 1940s there were three shafts at the colliery, No.1 and No.4 downcast shafts and No.2 upcast shaft.

The Swallow Wood Seam was worked from 1920 until it was abandoned in 1944. It was redeveloped in 1952 as part of the £6.5million Manvers Central Scheme. Under the reconstruction, carried out between 1950-1956, coal winding ceased at Kilnhurst. Following the reconstruction, coal from Kilnhurst was transported underground by diesel-hauled mine-cars, and wound at Manvers. Following reorganization, the No.4 downcast shaft was used for man-riding and materials, and the No.2 upcast for shaft inspection only.

On 1 January 1986 Kilnhurst was merged with Manvers to form part of the Manvers Complex, becoming known as Manvers Complex South. Things did not go well for the complex and, soon after, Manvers Complex South ceased production on 26 February 1988.

Seams Worked:

High Hazel (Kents Thick) Seam. 225yds deep, 2ft 4in to 2ft 6in thick, produced good house-coal.

Barnsley Seam, 283yds deep, produced from 1860 until exhaustion in 1942, 5ft to 5ft 6in thick, produced first quality coking coal, excellent steam and gas-coal.

Haigh Moor Seam, 359yds deep, 4ft to 4ft 8in thick, developed between 1930-1934 and again between 1939-1940, production was eventually discontinued due to extremely complicated haulage arrangements.

Parkgate Seam, 557yds deep, 1927-1966, 4ft to 4ft 9in thick. Due to geological disturbances near the shafts, the seam was very steeply inclined, varying from 1 in 3 to 1 in 4. As the workings advanced the gradient of the seam reduced. Produced first quality coking coal.

Silkstone Seam, 666yds deep, 1931 – 2ft 6in to 2ft 10in thick, first quality coking coal and high class house-coal.

Kiveton Park Colliery

Kiveton Park Collieries comprised two plants, Kiveton Park Colliery and West Kiveton Colliery. Kiveton Park Colliery was situated eleven miles south-east of Sheffield and nine miles north-east of Chesterfield. West Kiveton Colliery was some one and a quarter miles to the west of Kiveton Park. The Duke of Leeds was the principal lessor. In the mid-1890s there were some 6,000 acres of unworked coal, with an estimated 32m tons of reserves.

Kiveton Park Colliery was served by the Manchester, Sheffield & Lincolnshire Railway. The site of the colliery was chosen for its proximity to the railway connection between Sheffield, Grimsby and Hull, which provided an express route for shipping coal and receiving pit props and other materials, via the ports on the east coast. West Kiveton Colliery was served by the Killamarsh branch of the Midland Railway.

In March 1864 the Kiveton Park Colliery Co. Ltd was formed. The company comprised fourteen shareholders with an estimated capital of £200,000.

Sinking of the No.1 downcast and No.2 upcast shafts, both 13ft in diameter, known as the Kiveton Park Pits, commenced on 6 June 1866. Both shafts were sunk 401yds to the Barnsley Seam, which was reached on 6 December 1867. The shafts were 90yds apart and both were employed for winding coal. The first 100yds of each shaft were tubbed through the water bearing strata.

Hand got working of the Barnsley Seam, on the longwall system, commenced in 1868 and continued until late 1930, when mechanization was introduced in the form of coal-cutters and conveyors. However, cutter trials in the Barnsley Seam did not prove successful, and it was found that hand got output was greater.

In the mid-1890s there were thirty-two beehive ovens at Kiveton Park, utilising Parkgate small-coal. In 1870 the colliery began manufacturing its own gas, continuing production until 1926. The colliery office was built in 1872 and, as there was no school in the district, was utilized as a school for a time.

In 1874-75 two shafts were sunk to the Barnsley Seam at West Kiveton, some one and a quarter miles to the west of the Kiveton Park site. No.3 shaft was the 13ft diameter downcast and winding shaft, and the 13ft diameter No.4 shaft, 60yds to the west, was the upcast shaft, used solely for ventilation. Both shafts were sunk to the Barnsley Seam at a depth of about 260yds, and there were almost 100yds of cast iron tubbing in each shaft. Two large Guibal ventilating fans were located at the site, probably providing ventilation for the Kiveton Park site as well.

Kiveton Park Colliery, March 1985.

Coal from the Barnsley Seam, and later the High Hazel Seam, was raised at the West Kiveton shafts and screened there until 1897. After 1897 the shafts continued in use for ventilation and moving men and materials. Following the abandonment of workings in the West Kiveton area in 1931, the shafts at West Kiveton Colliery were filled.

About 1885 the two Kiveton Park shafts were deepened, the No.1 downcast shaft to the Thorncliffe Seam at 669yds, and the No.2 shaft to the Silkstone Seam at 733yds. The Thorncliffe Seam was worked to produce coking coal, but after a period of some seven years was abandoned as uneconomic. The coke ovens at the colliery ceased working soon after in May 1895.

Deepening of the two shafts in 1884 proved the following reserves:

Seam	Depth (yards)	Thickness (inches)	Tons (millions)
Clowne	209	44	31
Swallow Wood	454	36	28
Flockton	591	48	23
Thorncliffe	669	36	20

On 28 January 1900 working of the High Hazel Seam, at a depth of 310yds, was begun. The seam provided a good quality house-coal, but suffered from a weak, friable roof, which made it difficult to control. The seam was hand got until mechanization, in the form of coal-cutters and conveyors, was introduced in 1932. In November 1906 new screening plant was erected for the High Hazel Seam. In 1928 the colliery amalgamated with Sherwood Colliery, Nottinghamshire, for business reasons.

In 1930 the No.1 Barnsley pit bank was reconstructed and new screens erected, followed, in 1931, by the erection of a steel headgear. Pithead baths, with accommodation for 1,750 men, were opened in August 1938, together with a canteen.

In 1944 the colliery was taken over by the United Steel Co. and remained under the Company's control until nationalization. Power loading was finally introduced in 1950, when

Meco Moore cutter-loaders and trepanners were installed at the colliery.

In the early 1940s the colliery was working the Barnsley and High Hazels Seams. The Barnsley was hand got and the full section, varying from 5ft 1.5in, on the North Side, to 3ft 9.5in., on the South Side, was being worked. The High Hazels was mechanically undercut and loaded to conveyors, the full seam section of 4ft was worked. The output of the colliery was below the county average due to the 'dirty' nature of the High Hazels Seam, and the relatively high percentage of non-face workers employed underground. The age of the pit and position of the working faces having a significant influence. The royalty was heavily faulted, with dislocations running in north-west to south-east, and south-west to north-east directions, varying from a few feet to 120ft in size.

In 1957 Harrycrofts Shaft, previously part of another colliery, became part of Kiveton Park Colliery serving as a second ventilation shaft. In 1969 this shaft was abandoned and filled.

In August 1964 a major reconstruction scheme, costing in excess of £1million, was completed. At the surface a new coal preparation plant, to deal with coal from both the Haigh Moor and Barnsley Seams, was erected. Both seams were previously processed by separate screening plants. Electrification of the steam winders was carried out in 1963 – the No.2 in July and No.1 at the end of the year. In addition a new pit bank and headgear were constructed at the No.1 shaft and a new main ventilating fan installed. In 1962 the sidings were re-laid. Underground, a 374yd-long, 1 in 4.5 gradient, cross measure drift was driven between the High Hazel and Barnsley Seams, and a trunk belt installed in it to convey High Hazel coal to the Barnsley horizon. From this horizon High Hazel coal was hauled to the Barnsley pit-bottom, where it was wound to the surface, with Barnsley coal, in the No.1 shaft. The High Hazel shaft, which had wound men and materials for sixty years, was then used exclusively for man-riding. Known reserves at this time gave the colliery a life expectancy of seventy years. Later in 1964 the colliery suffered geological problems on some of the coalfaces, which resulted in a drop in output.

In late March 1971, production from the Barnsley Seam ceased. In 1972, after some seventy-two years of production, the High Hazel Seam was abandoned as uneconomic. In 1972-1973 the Clowne Seam was developed and extraction commenced, quickly replacing production from the Barnsley and High Hazels Seams. Initially the results from the Clowne Seam were disappointing due to the geological conditions, but with perseverance the colliery was once again made profitable.

A further reconstruction scheme, started in 1974, was completed in 1977, at a cost in excess of £2million, and the colliery changed from a shaft mine to become South Yorkshire's first full-scale drift mine. The Jubilee Drift, named to commemorate the Queen's Silver Jubilee in 1977, was 1,450yd-long, with a gradient of 1 in 5.8, sunk from the pit yard to link with existing Clowne Seam roadways some 200yds below the surface. In December 1977 the drift was officially inaugurated by the then NCB Chairman Sir Derek Ezra.

In the 1970s the colliery was producing high-quality industrial coal, most of which went to power stations, with the balance going to industrial and domestic consumers.

In the mid-1980s an £8.5million development took place in the Axle Area of the Clowne Seam, and by late 1987 the first face was producing coal.

By mid-1989 half a million pounds had been spent on the coal preparation plant in readiness for a planned merger. On 3 December 1989 Kiveton Park Colliery merged with nearby High Moor Colliery, and High Moor coal surfaced at Kiveton Park, where the coal preparation plant had been modified to handle the additional output. It was anticipated that the combined mine would have an output of 1.1m tons from the Clowne, Two Foot and High Hazel Seams.

The colliery worked for a few years more until it ceased coal production on 30 September 1994. Following closure the colliery was kept open on a 'care and maintenance' basis, before being offered, unsuccessfully, to the private sector. At closure the colliery employed 273 men and was supplying 6,000tons of coal a week to the Coalite plant at Bolsover, in Derbyshire. Kiveton Park was the last British Coal pit to use trepanner shearers in a seam less than one metre in thickness. On 19 October 1994, the *Daily Telegraph* reported:

Pits to shut
Two mothballed pits are to be closed after an unsuccessful attempt to find buyers, British Coal announced yesterday. Goldthorpe, near Doncaster, and Kiveton Park, near Rotherham, both South Yorks, will be sealed and demolished

Seams Worked:

Seam	**Depth (yards)**	**Thickness (inches)**
Clowne Seam	209	39
Two Foot Seam	254	25
High Hazel Seam	313	49
Barnsley Seam	401	57
Thorncliffe Seam	670	47

Manvers Main Collieries

Manvers Main Colliery consisted of two plants, No.1 and No.2 collieries, which were 650yds apart, about one mile from the two railway stations at Wath, and two miles west of Mexborough. The mineral property worked amounted to over 3,000 acres, part of which was the freehold of the Manvers Main Colliery Co. and the balance principally under lease from Earls Manvers and Fitzwilliam. The colliery was well served by railways, being located between the LNER and the LMS railways.

The partners in the new enterprise, which was known as 'The Manvers Main Colliery Co.' were: George C. Hague, Colliery proprietor of the Midland Ironworks, Rotherham; James Marrison; William Hunter; J.W. Hutchinson; Hilton Philipson; Henry Tennant and Alfred Allot. The company was registered on 21 February 1899 with a capital of £1,000,000.

The first sod of the colliery was turned on 21 May 1867 and the two pits were sunk to reach the 8ft 6in thick Barnsley Seam, at 284yds, in late 1870, when the first coal was raised. Both pits were 14ft in diameter. In 1873 the No.2 shaft was deepened to the Haigh Moor Seam, at a depth of 347yds. Its diameter between the Barnsley and Haigh Moor Seams increased from 14 to 16ft.

Between 1900 and 1904 two new shafts, No.3 and No.4, were sunk to the Parkgate Seam at a depth of 555yds. The No.3 shaft was 14ft diameter, and the No.4 shaft 16ft diameter and, like the No.1 and No.2 shafts, were located 650yds apart. The seams at Manvers had an inclination of 1 in 17 towards the east.

The arrangement of the shafts was such that the No.1 and No.4 were adjacent to each other, as were No.2 and No.3 shafts. The first 38yds of the No.1 and No.4 shafts were lined with cast iron tubbing, to hold back the feeders of water issuing from the Oaks rock, which amounted to almost 700gal. per minute. Similarly, No.2 and No.3 shafts were lined with 54yds of cast iron tubbing from the surface, to dam back the large quantity of water issuing from the Oaks Rock. At its peak the volume of water issuing reached 3,500gallons per minute.

By the early 1940s the No.4 shaft had become the common upcast shaft for all seams at the colliery. Though the shaft was fitted to wind coal from the Haigh Moor and Parkgate levels, it was only occasionally used to wind Haigh Moor coal. The No.1 shaft wound from the Barnsley Seam level, as well as Meltonfield and Winter coal from the Meltonfield level at 122yds. Winter coal was brought up a drift from the 150yd deep Winter level to the Meltonfield level, for winding to the surface. The No.2 shaft wound Haigh Moor coal from the Haigh Moor level at 347yds. The No.3 shaft wound Parkgate and Silkstone Seam coal from the Parkgate level at 555yds, the Silkstone coal brought up a drift from the Silkstone level at a depth of 648yds. In 1942 the No.2 winding shaft was deepened to the Swallow Wood Seam, and Silkstone Seam coal was wound in this shaft.

In about 1880 a battery of beehive coke ovens were built at the No.1 pit, and a further battery of coke ovens at the No.2 pit in about 1886. A brickyard attached to the colliery had a capacity of 120,000 bricks per week in the 1940s.

In 1888 the colliery company acquired limited status and became known as the Manvers Main Collieries Ltd. The company retained ownership of the colliery until nationalization.

In 1906 thirty-six Koppers Regenerative Ovens and thirty-six Simon Carves waste heat coke ovens were erected. The Koppers Ovens had a crude type of by-product plant. These were the first of a long line of coke ovens and by-product plant to be erected at the Manvers site. In 1912-13 the coke oven plant was further extended by the erection of fourteen Simon Carves waste heat ovens and a crude-benzol plant, followed by twenty-two similar ovens in 1917. In 1917 a benzol rectification plant was erected which manufactured high-grade benzol and toluol.

The production of gas from the coke ovens was of such a quantity that in 1930 gas was supplied to the Wath, Bolton & Thurnscoe Gas Co., and the Swinton & Mexborough Gas Co., as well as to

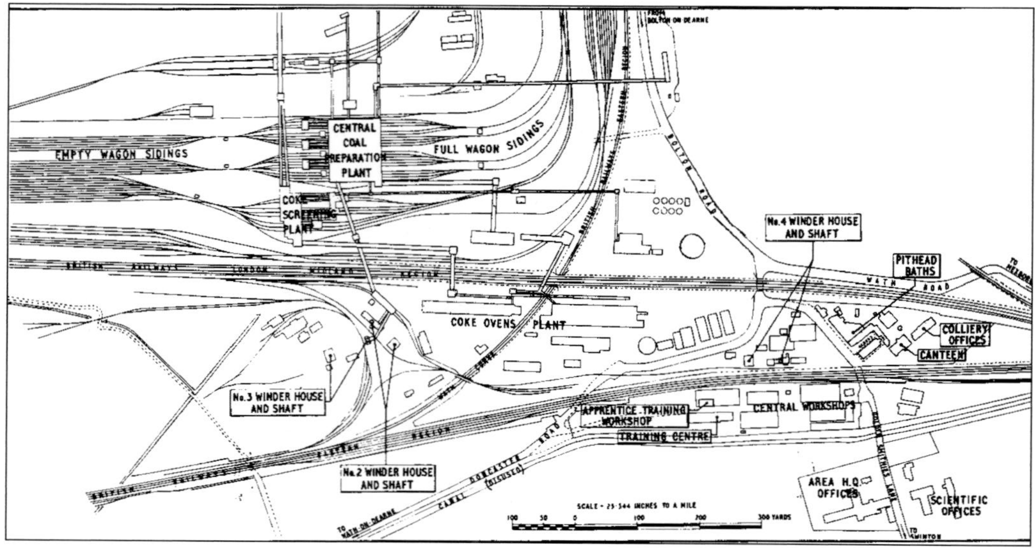

Manvers Main Colliery, plan of surface layout.

a number of large local companies such as Baker & Bessemer Ltd. In 1932 the decision was taken to dismantle the whole of the existing coke oven plant and erect a modern plant in its place. Following this decision, the old ovens were scrapped in November 1933 and a battery of thirty Otto underjet ovens erected. In June 1935 a further fifteen ovens were erected, making forty-five in total.

By the start of the 1930s sixty per cent of the output of the colliery was produced from the Barnsley Bed Seam, though the seam was nearing exhaustion after sixty years of extraction.

In the 1930s the main electrical generation plant was situated at the No.4 pit. The power station at Manvers was connected to the power station at Barnburgh via an underground 10,000V cable, and to the Yorkshire Electric Power Co. The system was further interconnected to the ring main of the Doncaster Collieries Association.

In May 1942 the Barnsley Seam, which had for so long been the mainstay of the colliery was exhausted and ceased production, and was replaced by the Haigh Moor Seam, which came into production in June 1942. The weekly target of the colliery at this time was 15,839tons. In 1942 the Cementation Co., of Doncaster, were preparing the No.2 shaft for coal winding from the Haigh Moor level. In the mid-1940s a second drift was sunk between the Parkgate and Silkstone Seams.

Manvers Combined Mine, a major redevelopment, was initiated in 1949, when a plan to concentrate the output from four collieries at Manvers was proposed. Called the Manvers Central Scheme, the plan was a major undertaking. The plan, to link four collieries and increase output by nearly 1M ton, was put before the Colliery Consultative Committees of Manvers Main, Wath Main, Kilnhurst and Barnburgh Collieries on Tuesday 18 October 1949. Combined mines had been introduced in continental Europe in the period between the two World Wars, but Manvers Combined Mine was to be the first in the UK.

The four collieries forming the proposed combined mine, though their ages varied from thirty-five to ninety years, still had immense reserves of very valuable coal. These reserves lay in what approximated to three horizons, each horizon comprising a series of associated seams, which could be worked simultaneously as a group.

Group I (upper seams): Newhill, Meltonfield & Winter Seams.
Group II (middle seams): Kents Thick, Barnsley, Dunsil & Haigh Moor Seams (moderately coking)
Group III (lower seams): Parkgate, Thorncliffe & Silkstone Seams (strongly coking).

The reserves in the three groups, in millions of tons, were as follows:

Group	**Barnburgh**	**Manvers**	**Kilnhurst**	**Wath**	**Total**
I	55.5	10.25	15.25	41.75	122.5
II	17.5	23.25	23	10.75	74.5
III	35	22	8.5	17.5	83
Total	**107.75**	**55.5**	**46.75**	**70**	**280**

On the basis of a daily saleable output of 12,900tons, the scheme had a projected life of ninety years. To trace the origins of the proposed project we need to go back to 1945, when Kilnhurst Colliery was acquired by Manvers Main Collieries. The surface facilities at Kilnhurst were generally in a poor state. The new, 20ft diameter, shaft was not equipped for winding, and was only used for ventilation. The previous owners, the Tinsley Park Colliery Co., had already started a scheme to use the new shaft to wind all the coal produced at the colliery, but had suspended the scheme. The main winding level for the future was to be at the Silkstone Seam level, at a depth of 644yds. By a stroke of good luck this depth was the same as the depth of the seam at the Manvers No.3 coal winding shaft. Examination of the joint plans revealed that conditions were favourable for transporting Kilnhurst coal to Manvers and winding it on the opposite shift to Manvers produced, Silkstone and Parkgate coal, at the No.3 shaft using skip-winding.

The plan was to construct a large central coal-preparation plant at Manvers, to deal with the combined output from the four collieries. Kilnhurst and Wath Collieries would be connected underground with Manvers, and Barnburgh Colliery would continue as it was, with the added improvement of concentrating the whole of its output at the No.6 shaft on two shifts. Barnburgh coal would then be conveyed by surface railway to the central coal-preparation plant at Manvers.

At the time of the plan, in the late 1940s, Manvers was working the Parkgate, Thorncliffe, Haigh Moor and Silkstone Seams and had estimated reserves of 55.5m tons of coal. For some years to come the output at Manvers would be produced from Groups II and III seams. Group II seams, at the time, were represented by the Haigh Moor Seam, skip wound at the No.2 shaft. Group III seams by the Parkgate and Silkstone Seams, wound in the No.3 shaft at the Parkgate level, Silkstone coal having been hauled up a 720yd, 1 in 20, drift to the Parkgate pit-bottom.

The reorganization programme was planned for the centralized working of coal reserves totalling 234.5m tons, and for the output of the four collieries comprising the scheme to increase from the 1948 figure of 2,314,855tons, to 3,225,000tons per annum. The scheme was originally estimated to cost £4million but eventually cost £6.5million. Manvers Main Colliery and the coal preparation and coking plants, built on land adjacent to the colliery, became the focal point for the scheme. Coal from Manvers, Wath and Kilnhurst collieries was wound from two horizons at Manvers Colliery, while coal from Barnburgh was wound at Barnburgh shaft, and brought overland to the complex by a rail link. The scheme concentrated coal drawing at three coal winding shafts, two at Manvers and one at Barnburgh. The three coal drawing shafts would be fully occupied for two shifts per day and would deal with the total output from the four collieries. The collieries would retain their separate identities for all purposes except coal winding and preparation,

Manvers Main No.4 man riding shaft, March 1988.

thereby disturbing the local loyalties of the workmen as little as possible. The scheme also eliminated the need for extensive reconstruction of the surface of two collieries, and made the maximum use of the coal winding and preparation facilities at the combined mine. The plan was to be completed by 1955 and envisaged developments over the following seventy-five years.

With the advent of nationalization, and the need to take advantage of all natural features, it was decided to make a new pit-bottom in the No.3 shaft, at a point 30yds below the Silkstone level. By early 1951 shaft deepening was underway. The pit-bottom was similar to that being constructed at Barnburgh Colliery, and an 8ton-capacity skip was installed in the shaft. The new pit-bottom was connected to Kilnhurst Colliery by a three mile long locomotive road, driven between the two collieries. The roadway was driven through solid ground, some 800yds of which was supported by roof bolts.

Both winding shafts at Manvers were equipped for skip-winding. The No.2 shaft wound about 4,500tons per day from the upper horizon at 343yds, in six ton-capacity skips, comprising 2,300tons from Wath and the balance from the Manvers Haigh Moor Seam. The lower winding horizon, at 670yds in No.3 shaft, wound 4,300tons, in 8ton-capacity skips. This comprised Manvers Silkstone and Parkgate coal, together with 1,400tons from Kilnhurst brought along the locomotive road between the collieries. The two shafts had a capacity of 8,400tons per day. The Wath and Kilnhurst shafts were then used for man-riding and the supply of materials.

In order to link Manvers and Wath Collieries, so that coal from Wath could be transported underground to be wound at Manvers, a locomotive roadway was driven in the stone between the two. Preparatory work at the Manvers end commenced in July 1949 and drifting began in February 1950. Preparatory work commenced at the Wath end in August 1951. Work continued intermittently, as the workers were used on other work during what was a complex phase of the scheme, and the two drifts were linked at 3.15a.m. on Friday 10 October 1952. The total length of the drift was 1,400yds, at a gradient of 1 in 400. When the work was underway the rate of advance was about 20yds a week. Other improvements included the construction of new pithead baths and canteen facilities at Wath, Barnburgh and Kilnhurst Collieries.

Coal from the combined mine was transported by belt conveyors to the Central Coal Preparation Plant. This plant comprised three identical units, arranged side-by-side, each having a capacity of 440tons per hour, giving a total of 1,320tons per hour. The three plants were designed to deal with; strong, medium and weakly coking coals, respectively and were arranged so that appropriate blending of small-coals could be achieved to enable the best mixtures for carbonization to be produced. The first unit of the plant came into operation in July 1955, the second in October 1955 and the third in June 1956. The plant was the largest coal preparation plant in Western Europe. It was served by twenty-four miles of railway sidings with excellent rail connections to the London, Midland and Eastern Regions of British Railways. The coke ovens at Manvers were supplied with coal direct by conveyor, and a central landsale depot was constructed to meet home coal requirements and supply local coal merchants.

At the Manvers Colliery site there were three batteries of Simon Carves coke ovens in operation at nationalization. In early 1952 the North Eastern Division of the NCB placed an order for the construction of coke ovens, by-product plant, and coal and coke handling equipment, to be erected at Manvers Main Colliery. The order was placed with Simon Carves of Stockport, Cheshire. The contract was for a battery of sixty-six coke ovens to be built on a new site, for the addition of twelve new ovens to the existing plant and for fourteen of the existing ovens to be rebuilt. Gas from the new battery of ovens, and twenty-seven of the existing ovens, was fed to the new by-product plant. The capacity of the coking plant was approximately 3,000tons of coal per day and was the NCB's largest installation

The new carbonization plant, situated close to the central coal preparation plant, was completed in 1956, at a cost of almost £9million. The new coking plant was a unit of the NCB Coal Products Division. Carbonizing 3,000tons of small-coal per day, the plant produced coke for the steel industry, benzol, sulphate of ammonia and a wide range of other products. Waste gas, produced by the plant, was supplied for local domestic consumption.

Further modernization was carried out in 1957, with the installation of new winders at the No.2 and No.3 shafts. These were the main coal drawing shafts for the colliery, and No.4 shaft was the man-riding shaft.

By 1959 the colliery had yielded approximately 65m tons of saleable coal, from the Barnsley Seam (exhausted by 1942) and the Haigh Moor, Parkgate and Silkstone (the seams worked at the time). The central coal preparation plant was fully operational and supplying the following markets; gas-coal thirty-four per cent, coke making thirty-one per cent, domestic fifteen per cent, industrial nine per cent, others eleven per cent.

By the early 1960s methane drained from Maltby, Thurcroft, Silverwood, Rossington, Yorkshire Main and Markham Main Collieries was being piped to the Manvers Carbonization plant. The aim was to use the gas to underfire the coke ovens, with the goal of eventually producing eight million cubic feet of coal gas. The advent of natural gas from the North Sea, during the mid-1960s, soon put paid to the scheme. By the 1980s the Silkstone and Swallow Wood Seams were being worked at the colliery.

In January 1986 Kilnhurst, Manvers and Wath Collieries merged to form Manvers Complex, with management control located at Manvers. Manvers became known as Manvers Complex Central, Kilnhurst as Manvers Complex South and Wath as Manvers Complex West.

The complex did not have an easy time and, in the five years between 1983 and 1988, had lost more than £54million. By the late 1980s a survival plan had been drawn up for the complex. The plan was to turn the colliery into a single face pit, worked from its central location, and to slim down the workforce to give the colliery a chance of becoming viable for the remainder of its limited life.

Manvers Main Nos 2 & 3 coal winding shafts, March 1988.

On 25 March 1988 the colliery closed. Soon after, the clearing of what was described as the largest piece of derelict land in Western Europe began.

Seams Worked:

Meltonfield Seam, 122yds deep, 3ft 4in thick. The seam suffered from water.
Winter Seam, 150yds deep, 2ft 5in thick.
Barnsley Seam, 284yds deep, 1870-May 1942, up to 8ft 6in thick of which 6ft 3in were worked.
Swallow Wood (Haigh Moor) Seam, 347yds deep, 4ft 6in thick.
Parkgate Seam, 555yds deep, 1901-1968, 5ft 3in thick.
Thorncliffe Seam, 641yds deep, 5ft 10in thick.
Silkstone Seam, 648yds deep, 2ft 11in thick.

Both the Barnsley and Parkgate Seams were 'handgot', though some mechanization was introduced into the Parkgate in its final years, during the extraction of pillars. Coal-cutting machines and belt conveyors were first introduced into the Meltonfield Seam.

Mitchell Main Colliery

Four miles south-east of Barnsley, on the western edge of Wombwell, the colliery was located on the side of the Dearne & Dove Canal Navigation and was served by the LNER.

Sinking of two 13ft 6in diameter shafts, the No.1 downcast and No.2 upcast shafts started in August 1871 and the Barnsley Bed Seam was won in September 1875 at a depth of 307yds. Sinking was undertaken by James Beaumont, Mining Engineer and Colliery Proprietor. Beaumont was also contractor for the New Oaks sinking in 1876.

Things cannot have gone well for the colliery, since it was offered for sale on 25 October 1882. In 1883 the Mitchell Main Colliery Co. was registered with a capital of £32,000. The company was to retain ownership for the colliery until nationalization.

In 1899 the Mitchell Main shafts were deepened to the Parkgate Seam, at a depth of 575yds. Following the abandonment of the Barnsley Seam, the decision was made to mine the seams below the Barnsley from Mitchell Main, and those above from Darfield Main Colliery. Mitchell Main concentrated on working the Fenton and Parkgate Seams and small areas of the Silkstone and Swallow Wood Seams. The Fenton and Parkgate Seams were mined as far as the Darfield Fault, the Parkgate over most of the take and the Fenton only in the western area. A connection in the Parkgate Seam was made to the No.3 upcast shaft at Darfield Main, which served as the upcast ventilation shaft for the eastern workings at Mitchell Main.

A report by Hugh F. Smithson, dated 19 February 1931, about the viability of Mitchell Main and Darfield Main Collieries, came to the conclusion that maintaining Darfiled Main was not a viable proposition. The report went on to recommend concentrating output at Mitchell Main, and abandoning Darfield Main except for pumping and ventilation, maintaining that this would yield considerable economic benefits. In the event it was Mitchell Main, which was to close twenty-five years later!

In the early 1940s the colliery was working the Fenton and Parkgate Seams. The coal from the Fenton Seam travelling down a drift between the two seams, which came out near the Parkgate Pit bottom, for winding from this level. Both shafts were equippied for coal winding, though the output was easily handled by the No.1 downcast shaft.

The Fenton was a difficult seam to work, because the roof of the seam was prone to collapsing. The high proportion of dirt to coal in the section worked resulted in the coal produced being very 'dirty'. On the west side the seam improved and the working section increased. The Parkgate Seam was also 'dirty' and suffered from poor roof conditions. In the early 1940s, one face in the Parkgate Seam, required the men to walk 2¾miles each way to their place of work.

Following nationalization, a further appraisal of Mitchell Main and Darfield Collieries was carried out, in the early 1950s, and the decision taken to close Mitchell Main. At Mitchell Main the poor surface installations, inadequate shaft ventilation for deep seam working, small diameter shafts and the expense of developing extensive workings in the working seams (Fenton and Parkgate), were not conducive to large-scale capital expenditure. In view of the long distances that coal was being hauled from the coalfaces and the exhaustion of the reserves it was decided to transfer the men to nearby Darfield Colliery and utilise the large No.3 shaft at Darfield to wind the output produced by the two pits. The manpower from Mitchell Main was predominantly transferred to nearby Darfield Main Colliery, swelling the workforce to 1,300.

When the workforce returned to work following the annual holiday in 1956, the *Sheffield Telegraph* of Tuesday 14 August 1956 reported on the closure as follows:

> ***New Pit Jobs but no upsets***
> ***Transferred to nearby colliery***
> *Returning to work yesterday after two weeks holiday with pay, 450 former employees of Mitchell Main Colliery, Wombwell, went to new jobs at neighbouring Darfield Main Colliery… The 'settling in' leaves about 50 men – mostly surface workers – temporarily displaced… Another 67 men have been left at Mitchell Main for salvage work and about 60 have obtained employment at other collieries on their own account.*

Mitchell Main closed in 1956, becoming a pumping station known as Mitchell Pumping Shafts. The make of water in the Meltonfield and Barnsley Seams necessitated constant pumping and pumps were located on both seam horizons for this purpose.

Darfield Main was reorganized and modernised at a cost of £700,000. The move went well and where possible the transferred men were given comparable work to that which they had at Mitchell Main. The Darfield branch of the NUM also agreed to pay death benefits to 150 retired Mitchell Main miners, and their dependents who also received free home coal.

Seams Worked:

Barnsley Seam, 307yds deep, 7ft 6in thick.
Fenton Seam, 550yds deep, 3ft 10in thick, the seam provided good coking coal.
Parkgate Seam, 575yds deep, 3ft 7in – 4ft 6in thick, the seam provided good coking coal.
Silkstone Seam, 690yds deep, 3ft thick.

Mitchell Main, early 1900s.

Monk Bretton Colliery

Monk Bretton Colliery, situated in a picturesque location two and a half miles north-east of the town of Barnsley, was one of the best laid-out and largest collieries in the Yorkshire Coalfield. The royalty of nearly 2,000 acres was leased from Viscount Halifax and others, to work the valuable coal-seams, including the Barnsley Bed Seam or 'Nine-foot' as it was known locally. The greater part of the land in the township of Monk Bretton belonged to the monks of Bretton, a member of the Cluniac order. Hence the origin of the name of both the town and colliery. Remains of the abbey still survive in the vicinity to this day.

The owners of the colliery were Mr T.W. Embleton, Mr W. Day, Mr W. Pepper and Mr T.M. Carter. Both Embleton and Day were engineers, Embleton was the principal mining engineer in Yorkshire and President of the Midland Institute of Mining Engineers. Day was an important colliery owner in the district. Under the direction of such auspicious owners it was not surprising that the colliery was so well laid-out. When complete, and in full working order, the colliery was designed for an output of about 1,600tons of coal per day.

The first sod of the colliery was cut in May 1867. Three shafts were sunk to the Barnsley Bed at a depth of 300yds, two 12ft-diameter downcast shafts and a 16ft-diameter upcast shaft. The downcast shafts were located 8yds apart and the upcast was placed 53yds from them. All three shafts were lined with cast iron tubbing to hold back the large feeders of water given off from the Lower Chevet Rock. The two downcast shafts were tubbed to a depth of 140yds from the surface, and the upcast shaft to a depth of 100yds. Below that an ordinary brick lining was set in hydraulic mortar, grouted behind with concrete. The upcast shaft was protected from the heat and acid, resulting from the underground furnace ventilator, by an additional 4in of firebrick casing.

The colliery was well served for distribution of its coal by the LMS Railway's Cudworth to Barnsley branch, running close to the colliery on one side, and the Barnsley branch of the Dearne & Dove Canal along the other. House-coal from the colliery had a good reputation and was distributed to London and the south of England. In addition, coal was distributed throughout Yorkshire by the Aire and Calder Canal. Hard steam-coals were used locally and were exported, principally through the port of Hull, having been transported by the Hull & Barnsley Railway.

The No.1 downcast shaft was the coal winding shaft, and the No.2 downcast shaft became the auxiliary, for lowering and raising men, timber, stores and materials.

Close to the road at the entry to the colliery were a number of substantially-built houses for the manager and other officials, together with a range of offices. A range of well-appointed workshops housed smiths, joiners and fitters. A wagon shed carried out the general repairs required at the colliery. Brick-making and drying machinery were provided, to use the 'bind' brought out of the pit to manufacture high-quality bricks. In the 1880s the colliery was capable of producing 18,000 bricks per day.

Workings in the Barnsley Bed were exhausted in 1939 when the colliery was purchased by the Barrow Barnsley Main Collieries Ltd, who retained ownership of the colliery until nationalization. Barrow Barnsley Main Collieries had purchased the colliery in order to work coal in the Upper Coal Measures, namely the Beamshaw, Kent's Thick (known locally as the Low Beamshaw) and Winter Seams. The Beamshaw Seam, at a depth of 182yds, was the first to be extracted by the new owners. The Kent's Thick, located about 40yds below the Beamshaw, was reached by a 100yd drift, dipping at a gradient of 1 in 3.3. All three seams were being extracted by 1946, and output was 1,500-1,600tons per day, and could be increased to about 1,850tons if required.

From July 1961, for the first time in the history of British mining, coal was being won on three shifts a day, while the waste was being simultaneously stowed into the gob on two shifts. On 10 February 1962 the Barnsley Chronical reported:

STOWING AT MONK BRETTON COLLIERY

At Monk Bretton Colliery, 900tons of coal are produced from one coal face – Kents 71S. every day, and 900tons of dirt are pneumatically stowed the same day into the void (or waste) left by the extracted coal. The face is mechanically cut, loaded and supported, and to make the machinery pay its way, coal is mined on three shifts throughout the twenty-four hours of the day.

When the face in the Kents Seam was planned it would be working at the relatively shallow depth of 310yds, beneath a plant built by Barnsley Corporation. To prevent subsidence damage it was necessary to stow the face void produced. As the face machinery cost over £100,000, it was decided that the face would be worked on a three-shift system and stowed simultaneously. The dirt for stowing came from the dirt extracted at the surface screening and washery plant, which was crushed and mixed with dirt from the spoil heap. The stowing material was dropped down pipes in the shaft and mechanically conveyed to the stowing machines near the coalface. The stowing machines then blew the dirt into the goaf. The scheme was well justified, as shown by the average face output per manshift of over seven tons, compared with four tons when the face was worked by conventional cyclic methods, and the negligible surface subsidence achieved.

Monk Bretton Colliery, with two downcast shafts and upcast shaft to right.

In the years up to closure in 1968, the colliery had made heavy losses, but in 1968 there was opposition to closure on the grounds that by developing a section in the Winter Seam, the workforce believed, the colliery could be made economic again. The opposition was unsuccessful and in April 1968 the colliery closed.

Seams Worked:

Seam	Depth (yards)	Thickness (inches)
Winter Seam	157	48
Beamshaw Seam	182	27
Kent's Thick Seam	222	44
Barnsley Seam	300	114

Orgreave Colliery

Located in the Rother Valley, the colliery was six miles east of Sheffield and three miles south of Rotherham. It was connected to the LNER to the west, and had its own branch line to the LMS to the east. When it closed in 1981 the colliery was the oldest working unit in South Yorkshire, having been in production for over 120 years. The colliery was located in an awkward position with a congested surface layout.

The two shafts, the No.1 downcast and No.2 upcast, were sunk to the Barnsley Seam in 1851. In 1870 the Fence Colliery Co. purchased Orgreave Colliery, which was a small undertaking at that time, from the Sorby family. At the time Orgreave Colliery was working the Barnsley Seam at a depth of 132yds. Between 1871 and 1872 production was suspended to allow major developments to take place at the colliery.

In about 1875 the Fence Colliery Co. was wound up and reformed as Rothervale Collieries Ltd. In 1918 Rothervale Collieries became a branch of the new United Steel Co. Between 1889 and 1890 the downcast shaft was deepened to the Silkstone Seam at 475yds.

In 1918, the Rothervale Colliery Co. built the Orgreave Coking Plant, adjacent to the colliery, to secure a market for the coal produced by their collieries. The plant, comprising forty-three Koppers compound regenerative coke ovens, later became part of British Steel Corporation, producing coke, gas, crude tar, sulphate of ammonia, and crude and refined benzole, as well as other products. The coking plant was to work until 1990, and figured prominently during the National Coal Strike of 1984-85.

In the early 1940s the colliery was working the Flockton, Parkgate and Silkstone Seams. The Flockton Seam varied between 3ft 4in and 3ft 10in in section, and the full seam section, as well as up to 8in of the roof, was taken. North of the shafts the Flockton Seam proved to be too dirty to work economically, and all the working faces were to the south-east and south-west. The Parkgate Seam was 3ft 7in thick, which was worked, together with 12in of the roof. The major problem with the Parkgate Seam was that it was subject to washouts. The Silkstone was 5ft 1in thick, including a 9in dirt band, and the full seam section, as well as 6in of the roof was worked. Some 1,700yds to the east of the shafts, the seam was subject to a thick dirt band, and beyond this point the coal was unworkable. In all three cases the soft material directly above the coal-seams was removed to prevent roof falls.

Orgreave Colliery, April 1983.

For safety reasons, the machinery in the Flockton & Silkstone Seams was powered by compressed air, while that in the Parkgate Seam was electrically operated. The Spa and Upper Whiston Faults, which occurred in the royalty, with downthrows of 120 and 80yds respectively, posed common problems for all the seams at the colliery. By the 1940s drifts had been driven through the faults so that they posed no further obstacle to future development at the colliery.

Following nationalization the colliery was reconstructed. The principal features of the reconstruction were the development of the Swallow Wood Seam, the introduction of skip-winding, installation of a new ventilation fan and the construction of new colliery offices.

The Parkgate Seam ceased production at the colliery in 1962, followed by the Silkstone Seam in 1965. The colliery was connected underground to Treeton Colliery in the Swallow Wood Seam, and was later sealed at the Treeton end, when Orgreave closed in 1981. In the 1970s the colliery supplied part of its output to the adjacent British Steel by-product plant, with the balance going to power stations and the domestic market.

Things were becoming difficult for Orgreave by the early 1970s, with the colliery facing geological difficulties. They were working the Swallow Wood and Flockton Seams at the time, and the Flockton was almost exhausted. In January 1973 the *Colliery Guardian* reported that, although the colliery was facing geological difficulties, it would not be closing. There was still a fair amount of coal left, though it would be difficult to mine. By the mid-1970s the Swallow Wood Seam was the last workable seam in the colliery take.

In October 1981 coal production ceased and the colliery closed, but the coal preparation plant on the Orgreave site was kept open in order to wash coal from nearby Treeton Colliery. In the late 1980s Orgreave coking works closed and was blown up on 16 November 1991.

Seams Worked:

Seam	Depth (yards)	Thickness (inches)
High Hazels Seam	39	40
Barnsley Bed Seam	132	96
Swallow Wood (Haigh Moor) Seam	202	57
Flockton Seam	328	39
Parkgate Seam	386	40
Silkstone Seam	475	48

Treeton Colliery

Treeton Colliery was situated three miles south of Rotherham, and was served by the LMS and the LNER. Located in a narrow valley the colliery had a constricted site, which was not conducive to expansion. Railway connections were made to the former Midland Railway, about half a mile distant, and to Orgreave Colliery, from where there was a connection to the former Great Central Railway.

Treeton Colliery was sunk by Rothervale Collieries Ltd, who leased 1,300 acres of minerals at Treeton, from the Duke of Norfolk in 1874. The special resolution authorizing the lease, also authorized the formation of Rother Vale Collieries Ltd, and the winding up of the Fence

Colliery Co. Ltd. The new company owned three collieries; Fence, Orgreave and Treeton, and was set up in about 1875, with a nominal capital of £300,000 and 2,865 acres of coal.

Cutting of the first sod of the colliery took place on Wednesday 13 October 1875, at which time there was a large gathering of ladies and gentlemen, as well as a considerable number of local inhabitants. Sinking of the two 13ft diameter shafts to the Barnsley Seam, at 333yds proceeded, and in 1876 the two shafts had reached a depth of 72 and 75yds. By 1877 the shafts had reached 323 and 288yds in depth. By this date the financial resources of the colliery were causing concern and an unsuccessful appeal was made to the shareholders. The response was disappointing and the company was left in debt to the bank, which was then left virtually in control of the company.

The Barnsley Seam was reached in February 1878, at a depth of 327yds, but due to the poor state of trading and a lack of capital, only a small area of coal was extracted before work at the colliery was suspended. Continuous production from the Barnsley Seam commenced in 1883, when the colliery produced 59,742tons, increasing to 187,346tons by the year ending 1886. Production from the High Hazel Seam commenced in 1888, the seam yielding a good house-coal.

The area worked by the colliery suffered from two major faults, with trends from north-west to south-east, which interrupted the workings. The Spa Fault resulted in a downthrow of 120yds and the Upper Whiston Fault of 80yds. The north-eastern boundary between the colliery and its neighbour Silverwood, was defined by the Kings Fault.

In 1898 a coal washery, of German origin, was erected at the colliery and proved to be a major improvement, working continuously until it was replaced in 1965. In 1910 and 1913 steam turbo generators were installed to provide electricity to the surface and pit-bottom.

As the colliery developed so the village of Treeton expanded, from a population of 312 in 1871, to 2,450 in 1901. A scheme was carried out for lighting the village streets by electric arc lamps, the colliery providing the power, Treeton village becoming the first village to have electric street lighting.

In about 1910 compressed air driven coal-cutting machines were introduced into the High Hazels Seam, working on longwall faces of about 100yds in length. In 1914 jigging conveyors, of German origin, were introduced for conveying coal along the longwall faces and discharging into tubs.

In March 1918 Rothervale Collieries was amalgamated with the newly formed United Steel Co. Ltd, who were to retain ownership of the collieries until nationalization. The aims of the company included the assurance of markets for the coal produced by the collieries owned by the company, as well as coke from the by-product plants, which were planned for construction at Orgreave and Thurcroft Collieries.

In 1930 a new screening plant and tippler house were constructed, and the surface railway track was reorganized. In 1932 a new electric winder, engine house and headgear were erected at the upcast shaft, and in 1937 both shafts were deepened to the Swallow Wood Seam at 404yds, and a new pit bank and headgear were erected at the downcast shaft. In 1943 a new electric winder and engine house were erected at the downcast shaft, followed by the demolition of the steam raising plant and chimneys in 1944.

In the early 1940s both the Barnsley and High Hazel Seams were being worked, using coal-cutting machinery. Electric powered machinery was employed in the High Hazel Seam and compressed air powered in the Barnsley Seam. The full seam sections, of 2ft 10in

Treeton Colliery with drift in foreground, July 1991.

in the High Hazel, and 4ft 6in in the Barnsley, were being worked. At this time, both seams were experiencing difficulties due to the Spa and Upper Whiston Faults. The Barnsley Seam workings to the east of the Upper Whiston Fault were in an extremely faulted area and a large pillar of coal, left to support Ulley Reservoir, interfered with workings in the High Hazel Seam. The weekly target in the early 1940s was 7,950tons.

In 1942 froth-flotation was added to the 1898 washery plant, to recover the significant quantity of small-coal which was being discarded.

In 1955 a trepanner coal-cutter was introduced into the High Hazels Seam. Initial trials proved very successful, but unpredictable geological conditions made working conditions difficult. The working of as many as eight seams of coal in the Treeton area was responsible for the formation of underground cavities. Water collected in these cavities and, on a number of occasions, broke through and flooded the workings at Treeton.

By the early 1960s both the Barnsley and High Hazel Seams, which had been the mainstay of the colliery since the early days, were nearing exhaustion. To compensate for this loss a major development scheme was started in the Wathwood Seam in 1963, and by 1966 this seam was producing the total output of the colliery. To further supplement this output, development of the Swallow Wood Seam was begun in 1972, to access 9m tons of coking coal, and by 1975 output was being taken from this seam. The Barnsley Seam ceased production in 1965 and the High Hazels in 1966.

A development scheme was completed at Treeton in 1966. A major objective of the scheme was to increase the capacity of the downcast shaft by installing 5ton-capacity skips. This, in effect, increased the potential capacity of the shaft by fifty per cent. To handle the increased output a new coal preparation plant was erected at the surface. In the financial year 1969-70 the colliery achieved the highest annual output in its history, when it produced 652,000tons.

Treeton was connected underground to Orgreave Colliery via the Swallow Wood Seam. This connection was sealed in 1981, following the closure of Orgreave. Treeton was also connected to Thurcroft Colliery, but via the Barnsley Seam. This connection was sealed in 1970 following an underground fire.

By 1975 the colliery had one customer, the Central Electricity Generating Board, to which it usually supplied two 900ton-capacity trains per day. Spoil from Treeton was being sent by rail to Orgreave Colliery, for dumping on the colliery tip, at the rate of 6,000tons per week. In 1975, after 100 years of mining, a calculation of the reserves remaining showed the following:

Seam	Reserves (millions of tons)
Wathwood	1.5
Barnsley	2.1
Swallow Wood	6
Haigh Moor	4.2

In 1976 a £20 million scheme was approved to transform the colliery and increase capacity. Work started on the construction of a 3,000m drift to reach the deeper reserves in the Swallow Wood Seam. Construction took from 1976 until 1981. The new scheme involved the transport of coal via a staple shaft and a shallow 816m tunnel, beneath the village of Treeton. By 1982 the system was working, and coal was flowing from the coalface to the coke ovens at Orgreave Coking Plant without seeing the light of day.

By mid-1987 the colliery was experiencing difficulties. Seismic and borehole exploration had revealed poor geology in a proposed area of reserves beyond the major Kings Fault. The only way for the colliery to continue on an economic basis, taking account of the fact that the colliery was now dependent upon reserves in the Swallow Wood Seam, was to reduce to single face working with a reduced workforce.

Productivity at the colliery had increased from 24.8cwt per man-shift in 1947 to 76cwt per man-shift in 1990, a three-fold increase.

The last day of coal production at the colliery was 7 December 1990. During its life the colliery is estimated to have mined 36m tons of coal and 10m tons of dirt. Following closure, salvage operations and work on underground water dams continued until 1991.

Seams Worked:

- Wathwood Seam, 181yds, 1963-1982
- High Hazel Seam, 2ft 10in thick, 243yds, 1888-1965
- Barnsley Seam, 4ft 6in thick, 333yds, 1883-1966
- Swallow Wood Seam, 4ft 2in thick, 406yds, worked around 1900-1947, and again in the period 1972-1990
- Haigh Moor Seam, 600yds (worked immediately prior to closure)

Waleswood Colliery

Waleswood Colliery was situated midway between the Woodhouse Junction and Kiveton Park stations on the Sheffield and Retford branch of the Great Central Railway. The colliery sidings connected to the main line of the Great Central Railway. Messrs Skinner & Holford, the proprietors of the colliery, leased considerable areas of mineral property in the area. The coal measures at the colliery had a general dip of 1 in 14 to the east.

Sinking commenced in about 1856, when two shafts were sunk to the 3ft 6in thick High Hazels Seam, at a depth of 97yds. The downcast shaft was 9ft 6in diameter. The upcast, located 30yds to the west, was 9ft in diameter. In 1864 two further shafts were sunk in close proximity to the original sinkings. The new shafts were sunk to the 4ft 9in thick Barnsley Seam, at a depth of 197yds. These shafts were a 12ft 6in diameter downcast, and a 9ft 6in diameter upcast shaft, located 180yds apart. The two upcast shafts from the later sinkings were some 20yds apart.

A small workshop and a blacksmiths at the colliery enabled general repairs to be carried out. Close to the colliery the company built a row of fifty workmens' cottages.

In 1883 a double-inlet Capell fan, the first to Rev. Capell's new design, was installed at Waleswood. The Capell was a very effective design of fan and rapidly became popular, being widely used throughout the coalfields of Britain and abroad. Many of the largest fans of the late nineteenth and early twentieth centuries were of this type. In 1897 a compressed air driven coal-cutter was introduced into the High Hazels Seam, and gave satisfactory results.

In 1903, during sinking to the lower seams, the decision was taken to dispense with pit ponies for underground haulage. The company found it difficult to get lads for pony drivers and were not in favour of the use of electricity, so the only alternative was compressed air. A system using compressed air powered winches was used to haul coal out of the headings. The system proved satisfactory and reduced haulage costs.

In the early 1940s the colliery was working the Parkgate, Thorncliffe and Flockton Seams, but output was well below the county average. Achievement of the weekly target of 6,200tons depended upon the success of the recently opened Parkgate Seam, but adverse roof conditions, caused by a thick dirt parting, which came down with the coal, prevented the target being achieved. Output was also restricted by the single coal-winding shaft, the upcast, which suffered from stoppages caused by the wooden cage guides. At the time, plans were in place to replace the rigid guides with wire rope guides. Two shafts were in use at the colliery at this time, the 384yd deep downcast to the Flockton Seam, and the 459yd deep upcast shaft to the Thorncliffe Seam.

Substantial losses were made over a number of years, both prior to vesting day and after. With the opportunity to work the remaining coal more efficiently at the adjoining Brookhouse (a mile to the north-west) and Kiveton Park ($1\frac{3}{4}$miles to the south-east) Collieries, it was decided to close Waleswood Colliery. In addition, both Brookhouse and Kiveton Park Collieries were profitable and formed an integral part of the Area Development Plan. Brookhouse had a projected life of sixty years and Kiveton Park of 100 years. At the time, all of the coal in the Thorncliffe Seam, with the exception of a small area more readily accessible to Kiveton Park Colliery, was exhausted. The remaining coal in the Parkgate Seam could only be worked with great difficulty, and it was decided to work this coal from Brookhouse. It was estimated that reorganization at Waleswood would cost £250,000, and even after such expenditure, the standard of efficiency would not be satisfactory. The remaining reserves amounted to 1M ton in the Parkgate Seam, and 860,000tons in the Thorncliffe, which at a weekly output of 5,000tons represented just seven and a half years of life. It was anticipated that the manpower at the colliery could be absorbed at local collieries.

After prolonged consultation with the NUM, the NCB set the closure date for 4 September 1948. The majority of the workforce, 636 out of almost 800, were offered employment at neighbouring collieries. A sit-in at the colliery resulted in nearly 100 men remaining underground as a

Waleswood Colliery, January 1979.

protest against closure. The strikers were served with one weeks notice and the NCB announced that because of the strike the pit would cease as a producing unit on Saturday 15 May 1948. At the same time a sympathy strike at Brookhouse Colliery was called off.

Seams Worked:

Seam	Depth (yards)	Thickness (inches)
High Hazels Seam	97	42
Barnsley Seam	197	57
Flockton Seam	370	39
Parkgate Seam	428	28in worked.
Thorncliffe Seam	459	50in worked of the 60in seam.

Wath Main Colliery

Wath Main Colliery was located between the Midland and the Great Central Railways, to which short branch lines were connected for the mineral traffic. Owned by the Wath Main Colliery Co., which retained ownership of the colliery until nationalization, the take of the colliery consisted of mineral properties amounting to 1,500 acres.

Sinking conditions were so wet that the first shaft was abandoned and sinking proceeded anew. Sinking was restarted and two new shafts, about 50yds apart, reached the Barnsley Seam in 1876, though the first coal was not wound until 1879. Both upcast and downcast shafts were 14ft in diameter and 346yds deep. The upcast shaft was to the north of the downcast shaft. Large feeders of water, amounting to about 3,400gallons per minute, were given out by the Meltonfield rock in the upper strata of the sinking, and two pumps were needed once the sinking had reached a depth of 45yds. Each shaft was lined with cast iron tubbing to a depth of 96yds.

A double row of sixty-four, 11ft diameter, beehive coke ovens at the colliery produced about 480tons of coke a week. A range of workshops served the colliery, providing equipment such as the Guibal type ventilating fan manufactured for use at the colliery.

The original timber headgear, erected in 1873 at the No.2 upcast shaft, survived until replaced by a steel headgear in 1970. This was one of the last timber headgears to survive at a British colliery.

Between 1912 and 1915 the two shafts were deepened to the 4ft 6in thick Parkgate Seam, at 609yds, and the first coal was raised in 1915. In 1923 the shafts were deepened to the 4ft thick Silkstone Seam at a depth of 689yds.

In 1923 Wath was one of the first collieries in Yorkshire to have pithead baths. These baths were in use until they were superseded by new baths and a canteen, costing £196,000, that were officially opened on 8 September 1962.

By the late 1920s the original winding-engine at the downcast shaft was no longer equal to the increasing duty placed upon it and it was decided to replace it. The constricted ground-space available favoured the installation of a vertical winding-engine. The first vertical steam winding-engine to be built for more than forty years was installed at Wath. An inverted vertical winding-engine, with the cylinders mounted above the winding drum, one of only three of this type of engine ever to be built in Britain, was installed at Wath in 1928. The engine, of massive design and construction, was built by Bradley & Craven of Wakefield. The engine was housed in a reinforced concrete building with external dimensions 40ft 4in by 42ft 8in, and a height of 53ft 6in above the foundations. The top of the engine house carried a 6ft deep water-tank. The engine house had three floors; the winding drum, post-brakes and brake engine on the lower floor, the control platform on the middle floor and the cylinders and throttle valve on the top floor.

In 1944 reserves in the Barnsley Seam were exhausted and the seam was abandoned, after sixty-five years of production. Soon after nationalization, rumours and fears that Wath Main was to close and its coal worked from other collieries, particularly Manvers Main, were rife. Senior officials met with the workforce and the rumours were denied. The local *South Yorkshire Times* of Saturday 22 February 1947, ran the headline: '**Wath Main NOT to close**'.

Prior to nationalization the owners of the colliery had planned a considerable modernization of the surface and the winding arrangements. After nationalization the proposal for a combined mine changed the priorities, and alternatives for getting Wath output to the Manvers No.2 shaft, in order to wind from the Haigh Moor Seam level, which had an unused winding shift available, were investigated. Several alternatives were investigated, and the decision was taken to drive an almost level locomotive road between the two collieries. Wath coal could then be skip wound on the opposite shift to Manvers Haigh Moor coal.

Between 1950 and 1956 Wath was one of four collieries (Manvers, Barnburgh, Kilnhurst, and Wath) which came together to form the Manvers Central Reconstruction Scheme. This £6.5million scheme enabled the combined output from four collieries, to be transported to a massive central coal preparation plant at Manvers. At Wath the principal feature of the scheme was the organization of the underground transport system. Coal was transported from the colliery's workings by diesel-hauled locomotive, along a two and a half mile locomotive road between Wath and Manvers. Following reorganization, the coal preparation plant at Wath was demolished and a new electric winding-engine house constructed. The estimated reserves at Wath totalled 24.5m tons and the plan was for coal from the Newhill, Meltonfield, Winter and

Wath Main Colliery, 1980s.

Kent's Thick Seams to be lowered to the level of the locomotive road by spiral chute and hauled to Manvers No.2 shaft, for winding to the surface.

In 1952 the Parkgate Seam was exhausted, followed by the Swallow Wood Seam in 1954. Between 1954 and 1957 the Meltonfield Seam provided the total output of the colliery, until the Newhill Seam was developed. Mechanization was introduced in the 1960s, with all faces in the Meltonfield Seam fully mechanized by 1965 and the Newhill Seam by 1973. The Newhill Seam was in production until December 1985 just prior to the merger with Manvers Colliery.

On 1 January 1986 the colliery was merged with Manvers Colliery. Manvers became known as Manvers Complex Central and Wath as Manvers Complex West. But things did not go well. Between April and December 1986 the Complex made a loss of almost £5million and by the end of the year was said to be losing £40 on every metric ton of coal produced. Despite having a further sixteen months of reserves, the men at Manvers Complex West decided to vote for closure, which came on 25 March 1988.

The area covered by the colliery has since been reclaimed and landscaped, making it difficult to visualise the colliery that once stood there.

Seams Worked:

Meltonfield Seam, 177yds deep, 4ft 2in thick.

Newhill Seam, 300yds deep, 5ft 6in thick.

Barnsley Bed Seam, 346yds deep, 1876-1944, 8ft 6in thick. A thickness of 6ft of the seam was worked, the 'softs' furnishing good quality house and coking coal, and the 'hards' excellent quality steam-coal.

Swallow Wood Seam, 408yds deep, 4ft 9in.
Parkgate Seam, 609yds deep, 1915-1952, 4ft 6in thick.
Thorncliffe Seam, 641yds deep, 5ft 10in thick.
Silkstone Seam, 697yds deep, 4ft thick.

Wharncliffe Woodmoor 1, 2 & 3 Colliery

Wharncliffe Woodmoor 1, 2 & 3 colliery was located two and a half miles north of Barnsley, and one and a quarter miles east of Staincross station, on the Great Central Railway line. A branch line was constructed from the colliery to the Great Central Railway, the colliery company's private line providing a connection with the Midland Railway at Carlton Sidings. A loading staithe on the Aire & Calder Canal provided a further outlet for output.

In the 1860s the Barnsley and Silkstone Seams were being extensively worked at the outcrop. Deeper workings to the Barnsley Seam were costly, which encouraged the profitable working of the better quality thin seams, such as the Woodmoor.

In September 1870 some 350 acres of the Woodmoor and Winter Seams at Carlton were leased to Joshua Willey, for a period of thirty years. The coal leased was under Lord Wharncliffe's estate. The colliery planned by Willey, New Willey Colliery, was to be quite small. Willey sank two 10ft diameter shafts 12yds apart, to the 3ft thick Woodmoor Seam at a depth of 40yds. The Woodmoor Seam was reached on 4 November 1871. The dip of the coal measures varied, but was generally 1 in 14 towards the east.

Soon afterwards the colliery was sold, and in 1873 a new colliery company, the Wharncliffe Woodmoor Coal Co., was formed. An additional lease, dated 16 June 1876, was granted to the Wharncliffe Woodmoor Coal Co. by the Earl of Wharncliffe. The new thirty-five year lease, ran from 1 February 1876. It was for three seams under about 350 acres, namely the Woodmoor Seam, the Two Foot Seam and the Winter Seam. The new company sank two new shafts to the Winter Seam at a depth of 72yds.

The new colliery comprised four pits: the No.1, or Winter Pit, was a 12ft diameter, 74yd deep, brick lined shaft; the No.2 Middle Seam (or Gas-coal) Pit was a 12ft diameter, 57yd deep, shaft; the No.3 Woodmoor Pit was 10ft in diameter and 42yds deep; and the No.4 Pumping Pit was a 10ft diameter shaft, sunk to the Woodmoor Seam.

The effect of the collapse of the coal boom in the mid-1870s was felt at the colliery. In November 1876 the colliery was put up for sale, to be sold as a going concern in one lot. The colliery was purchased at auction by Joshua Willey for £18,000, but he probably never worked it. In 1881 Howard Allport became the new owner of Wharncliffe Woodmoor Colliery. In 1883 Allport converted his privately owned colliery into a limited liability concern, which became known as the Wharncliffe Woodmoor Colliery Co. Ltd.

In 1888 the new shafts were deepened to the Kent's Thick Seam, at a depth of 137yds. The No.1 downcast shaft was used to raise Kent's Thick coal, production commencing in 1888. The No.2, upcast shaft, was located some 30yds south of the No.1 shaft. By the 1890s the colliery was working a mineral property of 1,430 acres, leased from Earl Wharncliffe. By the late 1890s there were 120 beehive ovens in operation at the colliery, producing almost 1,000tons of coke per week.

In the early 1940s five seams were being worked at the colliery, the Winter, Top Beamshaw, High Hazel, Low Haigh Moor and Lidgett Seams. The Winter and Top Beamshaw Seams were both machine cut and hand filled to tubs. Winter coal was transported down a cross-measures drift to the Beamshaw level, where it was wound from the Beamshaw horizon up the No.2 upcast shaft. High Hazels coal was machine cut, conveyed and wound up the No.1 downcast shaft. Lidgett coal was 100 per cent machine cut and conveyed up a cross-measure drift to the Low Haigh Moor level where it was raised up the No.3 downcast shaft together with Low Haigh Moor coal. Haigh Moor coal was fifty per cent machine cut and conveyed and fifty per cent hand got and filled to tubs. At this time there were fifty-three coke ovens, carbonizing 2,000tons per week, producing large furnace coke, crude benzol, concentrated ammonia and crude tar. A brickyard associated with the colliery produced some 50,000 bricks per week. The colliery was a victim of the closures of the 1960s, closing in August 1966.

Seams Worked:

Seam	Depth(yards)	Thickness (inches)
Winter Seam	77	40
Beamshaw Seam	90	36
High Hazel Seam	137	32
Kent's Thick Seam	157	38
Low Haigh Moor Seam	285	42
Lidgett Seam	321	28

Wharncliffe Woodmoor 4 & 5 Colliery

Wharncliffe Woodmoor Colliery shafts No.4 and No.5 were originally known as Carlton Main. Located three miles north-east of Barnsley, near Cudworth, the colliery was served by the LMS.

On 28 June 1873 about 8,000 acres of the Barnsley Bed Seam were leased to the Yorkshire & Derbyshire Coal & Iron Co. for a period of sixty years.

The Rt. Hon. Montague Stuart Granville, Earl of Wharncliffe, cut the first sod of the new colliery on Wednesday 12 November 1873. Two 14ft diameter shafts were sunk. The first shaft was completed in August 1874, and the second reached the Barnsley Seam, at a depth of 290yds, in July 1876. Sinking had proceeded rapidly, but heavy feeders of water, amounting to between 15,000 and 17,000gallons per hour, were encountered. To counteract the heavy feeders the shafts were tubbed, with what was the longest length of shaft tubbing to be installed at a colliery in South Yorkshire. The Barnsley Seam was expected to be 9ft in thickness and yield about 10,000tons per acre. It was proposed that the colliery would comprise upwards of 1,500 acres of coal.

In 1900 the Yorkshire & Derbyshire Coal & Iron Co. changed its name to the Carlton Main Colliery Co. In September 1909 Carlton Main was abandoned and closed in 1910. In about 1924 the Carlton Main shafts were taken over by the Wharncliffe Woodmoor Colliery Co. and were renamed Wharncliffe Woodmoor nos.4 & 5 shafts. In the early 1940s the Haigh Moor and Low Beamshaw Seams were being worked.

At Vesting Day the colliery was uneconomic and, to improve the viability of the colliery, a steep drift was driven from the surface to connect with the deepest workings of the mine. The drift had a vertical depth of 410yds and was probably one of the deepest drift mines in Britain. Coal flowed up the drift by belt conveyor to a new washery at the surface. Divisional output figures in 1957 showed the following output tonnage from the seams at the colliery:

Seam	**Tonnage**
Top Haigh Moor seam	30,000
Lidgett Seam	193,000
Beamshaw Seam	47,000
Winter Seam	167,000
Kent Seam	59,000
Total output	**496,000**

Following the closure of Wharncliffe Woodmoor 1, 2 and 3, in August 1966, it was feared that Wharncliffe Woodmoor 4 & 5 would follow quickly in its wake. However, Wharncliffe Woodmoor 4 & 5 worked on for a few more years, finally closing in July 1970.

Seams Worked:

Seam	**Depth (yards)**	**Thickness (inches)**
Winter Seam	155	52
Beamshaw Seam	178	27
Kents Thick Seam	212	28
Top Haigh Moor Seam	361	42
Lidgett Seam	–	–

Wharncliffe Woodmoor 4 & 5, early 1900s.

Wombwell Main Colliery

Situated close to the town of Wombwell, two and three quarter miles south-east of Barnsley, and ten miles north of Sheffield, the colliery was connected to the Manchester, Sheffield & Lincolnshire Railway and the Midland Railway. Hull and Grimsby were the principal ports used for the export of coal, to destinations such as the Baltic ports.

The first lease for the colliery was dated 19 November 1855. A provisional lease had been granted to the Darley Main Co., who undertook preliminary borings and exploration, before deciding to abandon the field as worthless, due to the 'throws' and the amount of water met with during the borings.

In August 1853 an agreement was made with the executors of Sir William Wombwell for coal in Wombwell at £250 for the first three years, £720 and £960 for the next two years and £1,200 thereafter. Sinking commenced on 8 December 1853, and reached the 224yd deep Barnsley Bed Seam on 28 October 1854.

The first lease was granted for the Barnsley Bed Seam, for ninety-nine years at £240 per acre. The colliery was subsequently converted into a private limited company under the name the Wombwell Main Co. Ltd, the Deed of Partnership dated 1 January 1855. The Wombwell Main Co. remained the owner of the colliery until nationalization. The aggregate take of the colliery amounted to some 1,300 acres, two-thirds of which belonged to the proprietors, the balance being leased. The take was remarkable in that there were no throws or faults in the entire area.

The No.1 south and No.2 north shafts were sunk to the 7ft 6in thick Barnsley Bed Seam, at 224yds 2ft. The two 12ft diameter shafts were placed 40yds apart, in a north-south line. The No.2 shaft wound coal from the Barnsley Seam. No tubbing was required, and the shafts were brick lined throughout. The water given out during sinking was unusually little in quantity, accounted for by the draining of the strata by the pumping operations of earlier workings to the dip. In 1862 the No.3 shaft was sunk to the Barnsley Seam, then to the Parkgate and further deepened to the Silkstone, at 587yds, in 1884. The No.3 shaft was located 70yds west of the No.1 shaft. This shaft was 12ft 6in diameter and became the upcast for the colliery.

Between April 1892 and April 1893 the No.1 shaft was deepened to the Parkgate Seam at 495yds. This shaft was 11ft 6in diameter to the Barnsley Seam, and 13ft diameter from the Barnsley to the Parkgate Seam. Development of the Parkgate Seam proceeded from Easter 1893, and working the seam by longwall cutting, the output averaged 1,200tons per day.
At the surface two ranges of coke ovens, eighty-six ovens in total, produced coke. The demand for coke at the time, exceeded output.

In 1862 both Wombwell Main and Wharncliffe Silkstone were on fire for much of the year, resulting in losses of about £24,000. Men 'in diving-bell dresses' were reported to be exploring the workings, to assess the state of the underground workings.

Wombwell steam-coal No.4 was placed on the Admiralty lists of the British and French Governments because of its high quality. 'Wombwell Hards', produced from the Barnsley Seam was regarded as superior Yorkshire coal. At the London Exhibition of 1862 Wombwell Main coals were awarded a medal.

In the early 1870s the colliery was capable of an output of 1,000tons per day, much of the output going to London and the South. In 1876 the Midland & Northern Coal & Iron Trades Gazette reported that, 'the best wages in the district are paid at Wombwell Main'.

By the late 1890s there were ninety beehive type coke ovens at the colliery, and the coke produced was consumed mainly at ironworks in Yorkshire and Nottinghamshire. In the early 1940s four seams were worked at the colliery, the Beamshaw, Fenton, Parkgate and Thorncliffe Seams. Beamshaw coal was wound at the No.2 downcast shaft. Fenton, Parkgate and Thorncliffe coal was wound at the No.1 downcast shaft from the Parkgate level, Fenton and Thorncliffe coal having been transported via cross-measure drifts to the Parkgate level. The last shift at Wombwell Main was worked on 23 May 1969, when the colliery closed.

Seams Worked:

Kents Thin Seam, 101yds deep, worked between 1966-69 but abandoned due to problems with thin sections of 20in.

Beamshaw Seam, 105yds deep, 2ft 4in worked.

Kents Thick Seam, 143yds deep, worked between 1966-69 but abandoned due to problems with thin sections 20-24in.

Barnsley Seam, 215yds deep, 7ft 6in to 10ft 6in thick, (Wombwell Hards was regarded as very superior South Yorkshire coal).

Fenton Seam, 480yds deep, 3ft 11in to 4ft 3in thick.

Parkgate Seam, 487yds deep, 4ft 6in thick.

Thorncliffe Seam, 520yds deep, 2ft 6in worked.

Silkstone Seam, 581yds deep, 3ft thick.

Wombwell Main Colliery, early 1900s.

Woolley Colliery

In the 1850s coal was worked on the hillside above Woolley Colliery, from adits driven into the shallow outcrop coal. In 1869 a shaft was sunk to the deeper seams at Wheatley Wood, about a mile from the site of Woolley Colliery proper. This first shaft was sunk to work the Winter and Beamshaw Seams. The shaft connected to roadways from the earlier adit workings, in order to provide through-ventilation.

In the early days the colliery comprised two plants, the Wheatley Wood (or Beamshaw plant) and the Barnsley Seam plant, which was located near to Darton railway station on the Lancashire & Yorkshire Railway, to which it connected with the branch line from Barnsley to Wakefield. The two plants were situated on the Woolley Estate, owned by Godfrey Hawksworth Wentworth, the mineral area totalling some 3,000 acres. The Barnsley Seam plant comprised a main slant (or inclined drift), some 1,800yds in length, driven in the Barnsley Seam. The Barnsley Seam was worked by longwall in two divisions, because of the thickness of the seam they took the lower coals first and worked the upper coals at a later date. Barnsley coal was brought to surface via the slant.

There were two shafts, placed close together, at the Wheatley Wood site. The 9ft diameter downcast shaft was 88yds deep to the Beamshaw Seam, and the 13ft diameter upcast shaft was 205yds deep to the Barnsley Seam. The Beamshaw Seam had a reputation as a high-class house-coal.

By 1878, nine years after sinking at the Wheatley Wood site, water seepage was becoming a serious problem and a Hathorn Davy differential pumping engine was installed to drain the workings. The pumping engine was placed at the bottom of the Wheatley Wood upcast shaft to raise water to an adit 140yds above the seam level. The adit was one and a quarter miles long and discharged into the river Dearne. Later in 1881 a Hawthorn Davey differential pumping engine was placed at the top of the upcast shaft, and also pumped to the adit. The steam pumps worked until the introduction of electric pumps. Water from the surrounding strata was a constant problem, and the Wheatley Wood shaft stayed in use for many years.

Two parallel north-south faults were proved on the east side of the property, one of which defined the boundary between Woolley and North Gawber Collieries.

In 1910-12 Fountain & Burnley, the owners of the colliery, sank two shafts at the site that became known as Woolley Colliery. Fountain & Burnley were to own the colliery until nationalization. The shafts were sunk to a depth of 423yds, intersecting the Parkgate, Thorncliffe, Silkstone and Blocking Seams. The first coal was drawn up the Woolley shafts in 1914.

The difficulty of working thin seams was evident and the viability of working the thinner seams at the colliery was questioned. A decision was made to fill the shafts to the Parkgate Seam level, but before this was executed there was a change of management and a vigorous programme of thin seam working commenced in 1915. Woolley Colliery had undertaken a great deal of pioneering work in mechanising thin seam working, gaining a nationwide reputation as a result.

In 1939 pithead baths were opened at the colliery. In the early 1940s the colliery was working the Parkgate, Thorncliffe, Silkstone and Blocking Seams. Coal from these seams was machine cut and conveyed for winding to the common winding level at the Silkstone Seam, at a depth of 321yds, where both shafts were used for coal winding. Parkgate coal was transported down a 1 in 5 cross-measure drift and Thorncliffe and Blocking coal, down 1 in 72 gradient drifts to the Silkstone level. In 1942 a third shaft was sunk, at the Woolley site, to the Lidgett Seam at 130yds.

In 1951 Roddymoor Colliery merged with Woolley. Following nationalization production at the colliery increased rapidly and it became necessary to improve the ventilation. As the complex workings of the colliery spread north-eastwards, the problems of ventilation grew. In the period 1959-1961 a large diameter ventilation shaft was sunk at the Wheatley Wood site, about one mile from the colliery, and a large capacity fan installed. This shaft was the deepest at the colliery at a depth of 613yds. This work together with other necessary reorganization at the colliery was expected to cost £3.5million.

By the 1960s the colliery was fully mechanized and a report in the *Sheffield Telegraph* of 15 December 1961 read:

ALMOST FULLY MECHANIZED

Pit looks forward to higher pay and output

Miners at one of Barnsley's biggest pits will soon be able to boast that their coal is untouched by hand from the moment that it leaves the coal face to the time it reaches the pit surface.

...To speed up production, skip-winding gear will also be introduced... Despite the mechanization, which will be completed by August next year, the pit will still maintain a link with the past for the colliery's forty-five underground haulage ponies will be retained to transport some of the lighter loads.

...The spokesman emphasized that the alterations would not result in any of the 2,100 workmen being declared redundant. The aim, he said, was for them to increase output and so enable them to earn high wages.

The colliery, reputed to have the longest roadway system in the coalfield, is estimated to have a future life of about 100 year. The roadway stretches 75 miles...

The colliery soon had two shafts, No.2 and No.3, fitted for skip-winding.

When the NCB instituted experiments in the remote operation of coalface equipment, Woolley Colliery was selected as the site for thin-seam trials of ROLF (Remotely Operated Longwall Face). Trials commenced in the Silkstone Seam on 28 of June 1965 and continued for a number of years, during which time much valuable experience was gained. On 3 February 1967 the *Colliery Guardian* reported, 'Woolley Boost', the ROLF4 face at Woolley was reporting a 100 per cent increase in productivity compared with conventional thin seam power loading. This was the world's first application of automated roof support in a thin seam, which in the case of ROLF4 at Woolley, was less than 30in thick. ROLF was a major breakthrough in automated mining technology, and opened the way for more seams to be worked profitably. The technique was successful, as shown by the increase in output, which reached 1,019,553tons in 1969.

In the 1970s almost the whole of the output from Woolley went to the industrial market, principally into the chemical industry and for the manufacture of coke for the steel industry. At this time the colliery was working the Top and Low Haigh Moor, Fenton, Thorncliffe, Silkstone and Blocking Seams.

In the period 1979-84 Woolley Colliery became the hub of the £102million 'West Side' Complex of the Barnsley Area, its central coal preparation plant receiving coal from a number of collieries up the West Side Drift, which was completed in 1980. At this time Woolley

Woolley Colliery, early 1980s.

Colliery was working the Kents Thick, Blocking and Whinmoor Seams, the Lidgett Seam was virtually exhausted by then. After a productive life of almost 120 years the colliery closed on 22 December 1987.

Seams Worked from the Wheatley Wood site:

Seam	Depth (yards)	Thickness (inches)
Winter Seam	69	39
Beamshaw Seam	97	31

Seams Worked from the Woolley site:

Seam	Depth (yards)	Thickness (inches)
Barnsley Seam	207	108
Low Haigh Moor Seam	–	28
Lidgett Seam	–	31
Parkgate Seam	240	32
Thorncliffe Seam	289	42
Silkstone Seam	313	31
Blocking Seam	342	25

8
Mining 1876-1900

The period 1876-1900 saw the consolidation of the central region of the coalfield and the sinking of large collieries further to the east, to work the Barnsley Bed Seam. Collieries developed during this period include Grimethorpe, Hickleton Main and Cadeby Main. These required large capital investment to sink and develop, and were consequently developed on a large scale for high output. These collieries produced up to and exceeding a million tons per annum, and employed 2,000 to 3,000 men. Cadeby Main Colliery, like its sister colliery Denaby Main before it, was now the most easterly and deepest colliery in Yorkshire, and Hickleton Main was to become, and remained for most of its life, one of the largest collieries in the country.

The typical size of royalties worked at this period varied from 3,000 to 5,000 acres, though in the case of the deep sinking at Cadeby Main, a royalty of 8,000 acres was worked. As well as the deep shaft mines sunk in the area covered by this phase, Ferrymoor Drift Mine was sunk within the confines of Grimethorpe Colliery, specifically to work the Shafton Seam.

Collieries sunk in this phase include:

Cadeby Main
Ferrymoor – situated within Grimethorpe Colliery.
Grimethorpe
Hemsworth
Hickleton Main
Highgate – sunk to work the shallow Shafton Seam.
Monckton Nos 1, 2 & 3
Rotherham Main
South Kirkby

Cadeby Main Colliery

Cadeby Colliery was half a mile south of the village of Cadeby, from which it took its name, and seven miles from both Doncaster and Sheffield. Occupying a cramped location at the entrance of the River Don gorge, between the river and the overlooking hillside, the colliery was one and a quarter miles from its sister colliery at Denaby Main, which was located on the opposite side of the River Don. Situated on the Great Central Branch Railway, running from Sheffield to Doncaster, Cadeby Colliery was served by the LNER and the LMS. Probably as a result of the cramped location, no subsidiary works were ever built on the colliery site.

Sinking of the shafts, located on the edge of the magnesium limestone, began when the first sod was dug on 25 March 1889. Following a protracted sinking, resulting from the immense volumes of water released during the sinking operation, Cadeby became the eastern-most and deepest colliery in Yorkshire. The two shafts were sunk to the south-east side of the Southerly Don Fault.

The extensive mineral property, amounting to some 8,000 acres, extended to the outskirts of Doncaster, in the east, and was held under lease by the Denaby & Cadeby Main Colliery Co. Ltd. The mineral property was a continuation of the workings at Denaby Main Colliery but was separated from them by a 70yd downthrow fault to the south.

Sinking of the two shafts commenced in March 1889 and was completed to the Barnsley Seam in February 1893. The No.1 Eastern, downcast, shaft was 16ft in diameter and 751yds deep. The No.2 Western, upcast, shaft was 16ft in diameter and 739yds deep. The two shafts were 87yds apart, and were both equipped for coal winding. A band of heavily watered strata, lying between 90 and 140yds from the surface, gave out heavy feeders of water, which flowed into the shafts at a rate of 5,000gallons per minute. In order to hold back the flow, both shafts were lined to a depth of 140yds, with a continuous length of cast iron tubbing.

A noteworthy feature was the Don Fault, which was close to the bottom of the shafts, displacing the strata by 126yds, with an upthrow to the north. The combination of this fault, and the declination of the coal measures to the south-east at three to four in. to the yard, resulted in the Barnsley Seam being 300yds deeper at Cadeby than at Denaby Main. Though the coal measures were generally of slight inclination, in parts of the take the dip increased to 1 in 3. A pillar of coal 300yds by 300yds was left to protect the shafts.

When sunk, Cadeby boasted the largest winding-engines in Yorkshire, manufactured by John Fowler & Co. of Leeds. Water for the boilers was drawn from the nearby River Don, and two boreholes provided Conisborough and Denaby Main with drinking water. The 180ft brick built boiler chimney was a landmark in the locality.

An extensive range of two-storey workshops, 280ft long and 34ft wide, was erected and fitted out with machinery, blacksmith's hearths and joinery shop equipment. There was also a tub repair shop and a foundry. Steam-driven DC generators provided current for electric lighting, both underground and on the surface, including the colliery offices, six houses and the managing director's house. The colliery sidings were very extensive, and laid out so that empty wagons gravitated towards the screens, and once filled, gravitated to the full sidings where they were made up into trains.

A 900ton-capacity Luhrig coal washer, erected in 1894, proved to be a 'white elephant' and was replaced by a Humboldt washer, with a capacity of 1,500tons per day, in 1898. Later a 1,875ton-capacity Baum type washer was added. All of these washers were electrically driven. A double row of 180 beehive coke ovens produced coke at the colliery, and at a later date ten patent coke ovens were added.

The Barnsley Seam was worked continuously from 1893 until the take was exhausted in 1966, providing the bulk of the colliery output. In the early days the seam was manually worked into tubs, later it was machine cut and loaded by hand, and finally, power loaded and hauled by diesel locomotives. Coal was conveyed from the colliery by both rail and the nearby River Don, which was navigable to the Humber. Coal was loaded at the staithes on the river and transported to Keadby and Hull. The first consignment of coal was despatched by rail on 14 January 1908, via the Great Central Railway.

In 1893 a new company, the Denaby & Cadeby Main Collieries Co. Ltd was formed. In March 1927 the company was amalgamated into the Yorkshire Amalgamated Collieries Ltd, which had a capital of £3.2million and comprised; Rossington Main, Dinnington Main, Maltby Main, Darton Main and the Strafford Colliery Co. Ltd.

In the 1920s the use of pulverized fuel was in its infancy, but Cadeby had an installation fired by pulverized fuel working from 1926. On 27 April 1936 the company was incorporated under the title the Amalgamated Denaby Collieries Co. Ltd, a title retained until formation of the NCB in 1947.

In the early 1940s the Barnsley and Parkgate Seams were being worked. Access to the Parkgate was via a pair of cross-measure drifts from the Barnsley Seam level. The Barnsley Seam had a section varying from 5ft 6in to 8ft 6in in thickness, of which between 5ft and 6ft 6in was worked. The seam was generally easy to work and the total output was hand got and loaded to tubs. Spontaneous combustion was a problem, particularly in the South District. The Parkgate Seam was 4ft 6in thick, and the full section was worked. In March 1948 the Hunslet Engine Co. supplied Cadeby with its first underground diesel locomotive.

Following nationalization, a reconstruction scheme was initiated to concentrate coal winding from Denaby and Cadeby Collieries at Cadeby No.2 shaft, using a clutch-drum electric winder to allow skip-winding from two horizons. The cost of the reconstruction was estimated at £4.3M, and was completed in early 1959.

In 1956 the steam winder at the No.2 shaft was replaced with an electric winder, operating skip-winding in the shaft. The skip was fed with coal from the Barnsley, Beamshaw, Dunsil and Parkgate Seams. On 1 November 1957 the No.1 shaft was closed, in order to replace the steam engine with an electric winder, and did not return to work until 9 February 1958. From this date until closure of the colliery all output was wound up the No. 2 shaft, No.1 shaft being used for materials and man-riding.

By 1956 Cadeby was winding all of the coal produced by Denaby Main. In May 1968 Cadeby and Denaby Collieries amalgamated and Denaby Colliery ceased to exist.

The decision to close the heavy-loss-making colliery was made on 7 November 1986. A combination of poor geological conditions, and no prospect of achieving better results, led to the decision. The colliery had not made a profit in the past twenty years, and following the 1983 review of the Beamshaw Seam, it was realized that the colliery had little life left. It was decided that the reserves in the Newhill Seam at Cadeby would in future form part of Barnburgh

Cadeby Main Colliery, 1979.

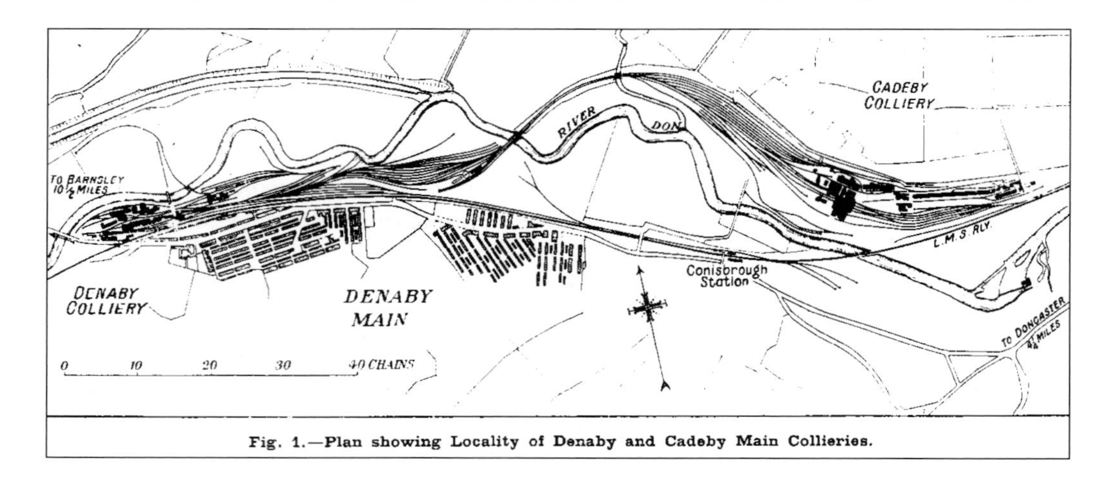

Fig. 1.—Plan showing Locality of Denaby and Cadeby Main Collieries.

Colliery reserves, and the reserves in the Shafton and Cudworth Seams would be worked by Goldthorpe Colliery. Of the 320 men at the pit at closure, 190 opted for redundancy and the remaining 130 were transferred to other pits, principally Rossington and Markham Main.

Seams Worked:

Beamshaw Seam, 632yds deep, worked between 1946 and 1955, and again between 1957 and 1966, 3ft 3in thick. When first worked the seam was cut and hand filled onto conveyors before loading into tubs, which were hauled by endless rope haulage to the pit-bottom. In the second phase of working modern coal-cutting machines and roof supports were used. The coal was power-loaded onto conveyors and loaded into diesel hauled tubs.

Barnsley Seam, 751yds deep, 1893-1966, 9ft 8in thick, furnished best quality house and steam-coal, and the fine coal was used for coke making. The seam was initially handworked into tubs, later machine cut and hand filled and eventually power loaded and hauled by diesel locomotive.

Dunsil Seam, 800yds deep, a small area of the Dunsil Seam was worked between 1953 and 1965, 6ft 4in thick.

Swallow Wood Seam, a small area was worked in the 1890s, though the main development came in 1960 and the seam soon constituted the total output of the colliery.

Parkgate Seam, 4ft 6in thick.

Ferrymoor Colliery

Originally sunk in 1915, probably by the Hodroyd Coal Co., the colliery was situated in Grimethorpe Colliery yard. The colliery worked the Shafton Seam at a depth of 56yds, the downcast shaft only used for coal drawing.

In 1919 the colliery was taken over by the Carlton Main Colliery Co. There followed considerable changes, including electrification of the colliery plant and the provision of electricity and compressed air from Grimethorpe Colliery. Electric power had originally been supplied by the

Central Power Station adjacent to Frickley Colliery. At the same time the screening plant at Brierley Colliery, about one and a half miles to the north, was dispensed with, and a surface tramway constructed between the two collieries, to bring Brierley coal to the enlarged screens at Ferrymoor.

Initially sunk as a shaft mine, the colliery became a drift mine in 1945 when a pair of surface drifts were driven, the No.1 drift reaching the Shafton Seam on 21 September 1945. Ferrymoor and Dearne Valley Collieries were adjacent to each other, working the Shafton Seam, and in consequence the two collieries worked closely together.

In June 1969 production started from Riddings Drift, a new drift with its surface entrance sunk in the pit yard of South Kirkby Colliery. The colliery was ventilated by the upcast at South Kirkby and shared the same surface facilities, but was operated as a completely separate unit. The drift worked the Shafton Seam by full retreat mining. The new drift was linked with Ferrymoor Colliery workings, to enable Ferrymoor coal to be worked and transported more economically.

Ferrymoor drift had been designed to meet the objectives set by the coal industry in the mid-1970s, of producing coal at highly competitive prices to compete with rival forms of energy. The project cost £611,000, and was designed to work known reserves in the Shafton Seam amounting to 7m tons, using retreat mining. The experience gained would prove to be valuable to the mining industry as a whole.

The drift employed 160 men, and in early 1970, they became the first British miners to produce coal at more than eight tons (162.5cwt) a shift for every man employed, this was almost a ton a man more than the previous record. The drift had a target of producing at a rate of ten tons a shift for every man employed. It was worked entirely by 'in seam' mining methods using the latest high powered machines and short production faces, which were extensively studied in the United States.

On 1 April 1973 Riddings Drift and the 300 man Ferrymoor Colliery, at Grimethorpe, merged to form Ferrymoor/Riddings Drift. The two collieries already shared common facilities and this was a natural progression.

Seams Worked:
Shafton Seam, 56yds deep, 54in thick.

Grimethorpe Colliery

Located in the parish of Brierley, the colliery lay four and a half miles north-east of Barnsley, ten miles west of Doncaster, and two miles north of the main Barnsley to Doncaster road, in a predominantly agricultural district. The mineral area leased was about 4,000 acres, initially leased by the Mitchell Main Coal Co.

The colliery was well served by both the London, Midland & Scottish and London & North Eastern regions of British Railways, and worked an area of coal of approximately seven square miles, later extended to some twelve square miles. The colliery royalty was relatively free from faults, with the exception of an area to the north-east of the take.

Grimethorpe Colliery was originally financed by the Mitchell Main Colliery Co. who commenced sinking of the No.1 downcast shaft in 1894. In October 1896, before completion of sinking in 1897, ownership was transferred to the Carlton Main Colliery Co. and the two companies were united under the name of the Carlton Main Colliery Co. Ltd in 1900. The Carlton Main Colliery Co. retained ownership of the colliery until nationalization. Sinking continued until the Barnsley Bed Seam was reached in 1897, at a depth of 560yds, and the finished depth of the shafts was 586yds. The No.2 shaft was then sunk to the Barnsley Bed and production commenced in 1897. Both shafts were 19ft in diameter, 80yds apart, and placed in a north-south direction. Barnsley Bed coal from the colliery was in great demand for steam-raising on land and at sea.

Considerable difficulties were experienced during sinking, due to the heavy feeders of water given out by the upper layers of strata, particularly at a depth of 75yds. To overcome these feeders the shafts were tubbed, but failures of the initial tubbing resulted in its removal and replacement. The upcast shaft was tubbed to a depth of 130yds and the downcast to a depth of 125yds. The No.1 downcast shaft was the main coal drawing shaft at the colliery. Adjacent to the colliery was a brickworks with a capacity, in the 1940s, of 200,000 bricks per week.

During the First World War a 14ft diameter staple shaft was sunk 272yds from the Barnsley Seam to the Parkgate Seam. At the same time the No.2 upcast shaft was deepened to the Parkgate Seam, at 854yds, to enable the seam to be worked. Openings were left in the staple pit for future working of the Haigh Moor and Fenton Seams, and the staple shaft was employed to ventilate the working seams. A winding-engine was located underground in order to wind from the Parkgate Seam to the Barnsley horizon and up the No.1 shaft to the surface. By this time the hand worked method of extraction had been replaced by the longwall system, employing conveyor belts. At the same time rope-haulage systems were replacing pit ponies as the principal means of haulage.

In 1925 sinking of the 21ft-diameter No.3 upcast shaft was commenced. In November 1927 Metropolitan-Vickers installed a 3,000-7,500 hp electric winder, the largest in the country at that time. The winder was capable of raising 300tons per hour from the 854yds deep Parkgate Seam. The No.3 shaft was originally intended to go down to the Parkgate Seam, but a recession in the coal trade resulted in the scheme being curtailed, and the shaft did not reach the Parkgate Seam until 1934. In the meantime it was decided to work the Beamshaw Seam, at 485yds, from this shaft, even though the sinking had proceeded some distance below the seam, and skip-winding was installed in the shaft. This was one of the first of such installations in the country. Between 1936 and 1940 skip-winding was employed on an experimental basis in the shaft. Beamshaw was the only coal raised in the No.3 shaft.

In the late 1930s a coal winding concentration scheme was formulated, using the No.1 shaft to wind the combined output from five coal-seams. The factors leading to this decision and the implementation of the scheme were as follows. Up to August 1939 the Beamshaw was wound in skips in the No.3 shaft, while Barnsley Bed, Haigh Moor and Parkgate coal was wound in tubs and cages in the No.1 shaft. In 1938 an extensive washout in the Haigh Moor Seam, was responsible for the decision to extend the workings at the colliery to the Melton-field Seam, via a pair of drifts from the Beamshaw Seam. However, it was not practical to wind the mixed output from the two seams by skip, as the Meltonfield was comparatively 'clean' in comparison with the coal from the 'dirty' Beamshaw Seam. It was also impracticable to wind Meltonfield coal by skip separately from Beamshaw coal or from No.1 & 2 shaft insets, as neither were equipped for coal winding. The decision was taken to dispense with the skip and wind coal from each seam using tub and cage. At this time the No.1 shaft was not working to capacity, due to reduced Haigh Moor and Barnsley Bed Seam tonnages, and implementing the scheme enabled coal from all five seams to be wound in the No.1 downcast shaft from the Barnsley level.

By the early 1940s the colliery was working, the Meltonfield, Beamshaw, Barnsley, Haigh Moor and Parkgate Seams. Coal from all five seams was wound up the No.1 downcast shaft from the Barnsley level. Meltonfield coal was taken down a cross-measures drift to the Beamshaw level, where it was taken with Beamshaw coal down the No.3 shaft to the Barnsley level. Haigh Moor coal came up a cross-measures drift to the Barnsley level, and Parkgate coal was raised up the staple between the Barnsley and Parkgate Seams. At the time the Meltonfield Seam was being developed. By the early 1940s the Barnsley Seam had been worked to its boundaries and had a limited life, at this time the full seam section of 5ft was being worked. The Haigh Moor Seam had a section of 3ft 4in and the full section was being worked. The deepest seam being worked, the Parkgate, had a section of 3ft to 3ft 6in and the full section was being worked. The colliery was fully mechanized by this date.

Following nationalization it was decided to group Grimethorpe and Houghton Main Collieries together. The scheme was a complicated one, involving reconstruction at both collieries, so that within the following fifteen to twenty years the combined output of the two collieries could be wound at Grimethorpe. The scheme cost £7.5million The two collieries were working in a large rectangular area of high grade coals. It was decided to reorganize and restructure the collieries so that Meltonfield, Winter and Beamshaw coal could be hauled by locomotive haulage to Grimethorpe No.1 shaft, for skip-winding to the surface. A locomotive roadway was to be driven to Houghton Main to collect Houghton Upper Series coals, for winding at Grimethorpe No.1 shaft. The No.3 shaft at Grimethorpe was to be deepened to the Parkgate/Silkstone series. Houghton Main No.3 shaft was to continue winding Parkgate series coals for an anticipated fifteen to twenty years, as required, until the output from the two collieries was raised at Grimethorpe, leaving Houghton Main No.1 and No.2 shafts for man-riding and supplies.

Grimethorpe Colliery, April 1993.

At nationalization, electric power was supplied to all the Carlton Main collieries, as well as Upton and Houghton Main, from the Carlton Main Co.'s power station at Frickley. The Frickley power station had been in operation since 1925 and was in need of major refurbishment. It was therefore decided to build a new power station at the Grimethorpe Colliery site, and to fuel the power station with coal from the large area of Kent's Thick Seam within the colliery's royalty. In 1958, at a cost of £5million, the new power station was commissioned. The station was visible for many miles because of its prominent cooling towers and 300ft chimney. In order to provide the special coal required for the power station, a new seam, the Kents Thick Seam, was developed. The power station was built to serve the electrification of collieries in the district and by 1959 was burning low-grade fuel and supplying power to collieries in the new No.4 Area.

In the same year the Beamshaw pit-bottom was reconstructed and locomotives were introduced into the eastern districts of the Upper Seams.

In 1953 a locomotive road, to join up the combined Grimethorpe/Houghton Main mine, was driven from the Beamshaw Seam pit-bottom to the East Melton Field. The roadway was known locally as the 'Golden Mile'. By late 1959 all west-side Meltonfield and Beamshaw coal was being brought along the roadway to the No.1 shaft at Grimethorpe, for skip-winding to the surface.

In 1954 the Winter Seam East District was opened and skip-winding was re-introduced into the No.1 shaft. By 1956 the Winter South-West District had been opened, and the locomotive road extended southwards into the Kent Thick Seam in order to provide fuel for the proposed power station. In 1957 the haulage in the east district of the Beamshaw Seam was connected via a drift to the Melton Field Seam.

In 1960 the No.3 shaft was deepened to the Thorncliffe Seam, at a depth of 894yds, and the ventilation was reversed. In 1961 a new coal preparation plant was completed. In the same year the Parkgate Seam was re-entered, and an underground connection made with Houghton Main Colliery, via the Kents Thick Seam, to enable Houghton Beamshaw coal to be raised at Grimethorpe No.1 shaft. In 1962 the No.2 shaft was relined between the Dunsil and Parkgate horizons, and deepened to the Thorncliffe Seam, at 894yds. In the same year skip-winding was reintroduced into the No.3 shaft, winding from the Thorncliffe horizon.

In 1963 a further underground connection was made with Houghton Main Colliery, this time at the Parkgate Seam horizon. By 1963 the combined mine reconstruction, enabling Houghton output to be wound at Grimethorpe, was complete. In 1966 a Coalite plant for the production of manufactured fuels was completed.

By 1972 the 5ft thick Fenton Seam was worked for the first time at the colliery, and was helping the pit to produce an output of just over two and a half tons per man-shift, for the first time in its history. In 1979 the drivage of a surface drift to the Fenton Seam was begun.

In the late 1970s two thirds of the joint output at Grimethorpe was supplied to produce electricity, twenty per cent for coking and ten per cent for the production of manufactured fuels, produced by the Coalite plant.

In September 1979 the reserves in seams being worked were estimated as follows:

Newhill, 7.5m metric tons
Melton Field, 4.5m metric tons
Low Beamshaw, 3.1m metric tons
Fenton, 9.2m metric tons

Parkgate, 8m metric tons
Total 32.3m metric tons

In addition the following reserves were also available in seams not being worked:
Thorncliffe, 10.2m metric tons
Winter, 6.1m metric tons
Kent Thick, 3.4m metric tons
Top Haighmoor, 3.2m metric tons
Top Silkstone, 12.9m metric tons
Total 35.8m metric tons

In the period 1979-84 Grimethorpe Colliery became the hub of the £174 million 'South Side' project complex of the Barnsley Area. Barnsley Main, Dearne Valley, Houghton Main and Grimethorpe were linked underground, the output from these collieries surfacing at Grimethorpe up a new one and a half miles tunnel, completed in 1980, for processing in the central coal preparation plant. On 22 April 1985 the first coal was brought up the South Side Drift from the Fenton Seam at Barrow Colliery. The drift was planned for an eventual capacity of 3m metric tons. Almost nine miles of new roadway had been driven underground to connect the four collieries. The South Side coal preparation plant was one of the world's most advanced and had cost some £28million. The preparation plant could blend more types of coal than any other plant in the UK, taking coal from five different seams, and achieved a record 40,790 metric tons daily output and weekly output of 160,000 metric tons. In the financial year 1980-81 the colliery produced its highest output of 1,225,486 metric tons.

In the late-1980s the colliery was working the Newhill, Fenton and Parkgate Seams, with a workforce of 900 men, and a planned production target of 1m metric tons in 1990. In the week ending 5 November 1988 the colliery achieved its highest weekly output of 34,346 metric tons, and during the year produced coal at a rate of five metric tons per man-shift, a colliery record. In October 1992 British Coal announced thirty-one closures with the loss of 30,000 jobs, with ten collieries earmarked for immediate closure. Grimethorpe, despite having made a profit of £700,000 in the year ending 31 March 1992, was one of the ten collieries.

On 30 October 1992 the colliery ceased production, and in September 1993 was one of the collieries offered to the private sector, but which did not receive any bids to continue mining. As a result British Coal moved swiftly to close the colliery.

The colliery boasted one of the foremost brass bands in the country, the Grimethorpe Colliery Band. Nationally known, and a major part of the community life of the village, the band was often heard at outdoor functions in the local parks.

Seams Worked:
Newhill Seam, comprising two beds of coal separated by a dirt band the seam had an overall thickness of 8ft 7in.
Melton Field Seam, 421yds deep.
Winter Seam, 440yds deep.
Beamshaw Seam, 463yds deep, a single bed of high-quality coal some 36in in thickness, producing a medium coking coal.

Kents Thick Seam, 494yds deep.
Barnsley Bed, 567yds deep, the seam was some 7ft thick.
Top Haigh Moor Seam, 668yds deep.
Fenton Seam, 809yds deep, a seam comprising three beds of coal separated by dirt bands and averaging 7ft 2in in total thickness. Methane drained from this seam was piped to the surface via the No.3 shaft, from where it was piped to a near by brickworks to be utilized for kiln firing.
Parkgate Seam, 839yds, comprising two beds of coal separated by a dirt band the seam had an overall thickness of 6ft 10in. The seam produced a strongly caking coal.
Thorncliffe Seam, 894yds deep.

Hemsworth Colliery

Located at Fitzwilliam, in the parish of Hemsworth, seven miles north-east of Barnsley, the colliery was on the northern fringe of the South Yorkshire Coalfield. The colliery was close to the Leeds & Doncaster Railway, which was jointly owned by the Great Northern Railway and Great Central Railway. The mineral property worked by the colliery was owned by Earl Fitzwilliam and several other proprietors, from whom the colliery leased some 3,000 acres, calculated to contain about 75m tons of coal.

Trial borings at Hemsworth reached the Shafton Seam at a depth of 100yds. Following the successful borings on Earl Fitzwilliam's estate at Kinsley, sinking at Hemsworth began in 1878. Until the end of December 1894 the only seam worked was the Shafton.

There were three shafts, Nos 1, 2 and 3, at the colliery. Shafts Nos 1 and 2 were sunk to the Shafton Seam, at a depth of 143yds, sinking beginning in 1878. No.1 shaft was 11ft in diameter and was the pumping shaft. Shaft No.2, located 50yds east of No.1, was 14ft in diameter and was an upcast shaft. To prove the lower seam the No.2 shaft was deepened, between 1897 and 1898, to the Haigh Moor Seam, at a depth of 685yds. In sinking to the Haigh Moor, the Barnsley Seam was proved at a depth of 612yds. By the late 1890s the Haigh Moor Seam was being worked from the No.2 shaft.

In 1893 the No.3 shaft, located 19yds east of the No.2 shaft, was sunk to the Shafton Seam, with a diameter of 18ft. In 1898 the No.3 shaft was deepened to the Barnsley Seam, at a depth of 612yds. The No.3 was a downcast, and coal from the Barnsley and Shafton Seams was raised in this shaft.

By the late 1890s three seams were being worked, the Shafton, Barnsley and Haigh Moor.

The colliery was not successful and in the late nineteenth century was sold to the New Hemsworth Colliery Co. In 1903 the colliery was sold to Fitzwilliam Hemsworth Collieries Ltd, and sold again in 1905 to South Kirkby, Featherstone & Hemsworth Collieries Ltd, who owned the colliery until nationalization.

Conditions were difficult for the company, and a dispute between Fitzwilliam Hemsworth Collieries and its workforce in August 1905, resulted in a strike and the eviction of the striking miners from their homes. The strike had been caused by a disputed price list governing payments for work done by the workforce. The company was in financial difficulties and refused to abide by the decision arrived at through arbitration, which resulted in the strike.

Following nationalization the principal concern about Hemsworth Colliery was the small

Hemsworth Colliery, early 1900s.

diameter of the No.1 and No.2 shafts.

Until shortly after nationalization, Hemsworth had been a hand-won colliery. Soon after, an entry was made into the Low Haigh Moor Seam and a single mechanized face installed. The face was installed to gain experience of working a mechanized conveyor face, and ultimately, to replace output from the Top Haigh Moor Seam workings, which were a long way out from the shafts. 'The future of the Low Haigh Moor Seam depends on the success achieved with this face', reported the manager Mr L.J.Mills '...and the future layout of the pit also depends upon it...' The Barnsley and Top Haigh Moor Seams were extensively worked at the colliery until 1964, when the Top Haigh Moor Seam was closed, and the remaining coal from the Barnsley Seam was conveyed underground to be wound at South Kirkby Colliery.

In July 1967 Hemsworth Colliery was merged with South Kirkby Colliery and the coal produced at Hemsworth was wound at South Kirkby. The two collieries had been physically linked underground for many years, and this was a natural progression.

Seams Worked:

Seam	Depth	Thickness
Shafton Seam	143yds	3ft 9in
Barnsley Seam	613yds	8ft 9in
Dunsil Seam		
Haigh Moor Seam	685yds	3ft 6in

Hickleton Main Colliery

Hickleton Main Colliery was situated in Thurnscoe, midway between Barnsley and Doncaster. The proprietors of the colliery were the Hickleton Main Coal Co. Ltd. Hickleton Colliery was one of the deep, expensive collieries sunk towards the close of the nineteenth century. An agreement was made between the Hickleton Main Colliery Co., the Earl of Halifax and Rev. Thornley Taylor, to work the coal beneath their land. The colliery was sunk to work the Barnsley Seam and was to become, and remain for most of its life, one of the largest in the country.

The LMS Railway and the LNER, both of which ran into the pit yard, served the colliery. Brickworks adjacent to the colliery were the only subsidiary works.

The sinking of two 18ft diameter shafts, the No.1 downcast and No.2 upcast shafts, was started in December 1892 and completed on 28 June 1894, to the base of the Barnsley Seam, at 537yds. The two shafts were placed in an east-west direction and were 56yds apart. During sinking a 30yd thick bed of sandstone was encountered at a depth of 19yds. The 'jointed' nature of the bed liberated large feeders of water of up to 1,000gal. per minute, which were eventually stemmed by the insertion of 50yds of cast iron tubbing at a depth of 7yds from the surface.

The No.1 pit-bottom was a remarkable piece of colliery engineering, being large enough to accommodate a pair of double-deck cages and hydraulic loading and unloading apparatus. The structure took sixteen months to build and required more than half a million bricks, with white glazed bricks on the face of the arch. The pit-bottom also included pit offices and stables. The employment of hydraulic loading and unloading plant and later, the use of mechanical stokers on the boiler plant, was very advanced for a late nineteenth century colliery. The colliery also had extensive sidings, claimed to be the best laid out in Yorkshire, if not England.

Working of the Barnsley Seam proceeded rapidly and in 1895 the colliery was producing 500,000tons, and by 1898 was employing 900 men and boys underground and 200 on the surface. On 17 August 1896 a supper was held to celebrate the raising of 6,000tons in one week, and by February 1899 the pit was beating all records.

Electric motor and compressed air powered forms of haulage, as well as pony haulage were used underground and electric lighting was used, both on the surface and underground.

The No.1 headgear was a steel headgear manufactured by Davy Bros, Sheffield, and the No.2 was manufactured from timber. On 27 July 1898 the timber headgear caught fire and the flames were visible for miles. The replacement steel headgear was manufactured by the Grange Iron Co.

Workshops, comprising fitting shop, blacksmiths, joiners and a saw-mill, and stables were provided in a building measuring 305ft by 38ft. Shaking screens and picking bands were in operation at the colliery and in 1898 a coal washer, of Wood & Burnett's design, was being erected by the Grange Iron Co. of Durham. When complete the washer had a throughput capacity of about 1,000tons in nine hours. In 1898 a double row of beehive coke ovens was under construction.

The colliery was financially successful and during the period of the First World War the colliery was returning dividends in the order of ten to fifteen per cent.

In 1923 the No.2 shaft was deepened to the Thorncliffe Seam, at a depth of 827yds, and a third shaft, No.3, was sunk to 908yds. The lower 74yds of the 21ft 6in diameter No.3 shaft, sunk between 1921-23, was filled to a point just below the Thorncliffe Seam. New screening plant was built to handle the increased output, and the No.3 shaft replaced the No.2 as the upcast for all seams at the colliery.

Between 1894 and the deepening of the No.2 shaft in 1923, the Barnsley Seam was worked ex-clusively, being wound at the No.1 and No.2 shafts. In 1910 the colliery raised

1,001,355tons at the No.1 shaft alone, which must be a record for a single shaft at the time. Barnsley Seam workings continued until May 1956, when the last of the coal in the North-East Plane Pillars was being worked.

The early system of working was by longwall faces, divided into tub stalls from which the coal was hand got and loaded into tubs. Transport to the pit-bottom was provided by endless-rope haulage. In the mid-1930s, following the introduction of rubber-faced conveyors, longwall face working took place, with the coal machine-undercut, fired, and hand filled onto conveyors. In the early 1950s power loading was introduced.

Development of the Parkgate Seam commenced in 1925 from the No.2 and No.3 shafts. The colliery continued to be financially successful, returning a dividend of ten to fourteen per cent in the late 1930s and early 1940s.

In 1937 the colliery joined the newly formed Doncaster Amalgamated Collieries Ltd, who retained ownership until nationalization.

The Newhill and Meltonfield Seams, which were above the Barnsley Seam, were first entered in 1936. In January 1950 the NCB granted an interim authorization to allow the two seams to be developed over a large area by the use of underground locomotives and mine-car haulage. The Meltonfield Seam, which was intersected at a depth of 389yds, was developed from No.2 and No.3 shafts from 1940, and wound from the No.2 shaft until September 1961. The whole seam section was generally worked, except in a few areas where 6-9in of top coal was left to support the roof. The Newhill Seam was developed from drifts driven from the North-East Plane of the Melton Seam in 1942. The Newhill workings were suspended after a year, but were resumed in 1950 and worked until July 1961. A washout in the seam was encountered approximately 1,100yds north of the shafts.

In the early 1940s the was working the Barnsley and Parkgate Seams, and development had just started in the Dunsil Seam. Each of these seams was affected by a series of faults running in a north-west to south-east direction. In addition, the Parkgate Seam suffered from large irregular washouts and had a poor roof over large areas, which made roof support difficult and resulted in a lot of 'dirty' coal being filled. Between 5-8ft of the 8ft thick Barnsley Seam was worked. The full seam section of the Parkgate and Dunsil Seams were worked. Barnsley and Dunsil coal was wound in the No.1 downcast shaft, and Parkgate coal in the No.3 upcast shaft. The No.2 downcast shaft was equipped for coal winding, but was principally used for man winding.

The Dunsil Seam was worked in two separate areas, the North-East and East Areas. The North-East Area was worked from drifts driven from the North-East Plane and Return between 1942 and 1948. The East Area was worked by drifts from the East Plane and Return, from 1946 until closure of the seam in September 1960. In both cases the full seam section was worked.

In 1955 the NCB authorized a reconstruction scheme, incorporating the interim authorization granted in 1950. The reconstruction was extended to include shaft winding and mineral handling reorganization, new pit-bottoms to serve the Newhill and Meltonfield Seams, combined at one level, and increased electrification to provide power to the two new electric winders at No.1 and No.2 shafts. The scheme was designed to replace output from the Barnsley and Dunsil Seams by increasing output from the Newhill and Meltonfield Seams and improving the efficiency of the mine to achieve an output of about 1.25m tons by 1962. The scheme also replaced old, worn-out plant and improved the surface layout at the colliery.

Between 1961 and 1962 a major reconstruction of the surface took place. It comprised: new winding-engines, to replace steam winders at No.1 and No.2 shafts; skip-winding at the No.2 shaft; new headgear at No.1 and No.2 shafts; a new mine-car system at the No.1 shaft; a new screening plant and a 750ton bunker, together with further underground mechanization. Dirt disposal arrangements by motor dumper trucks, was introduced to replace the old aerial flight. In September 1961 a reconstruction scheme for the working of the Meltonfield and Newhill Seams was completed. This took the form of a horizontal drift from the No.1 shaft, at a depth of 463yds. Coal from the Meltonfield Seam was loaded into 3ton mine-cars and hauled along the drift by diesel locomotives, to be drawn up the No.1 shaft in cages. A 2,000hp ground mounted koepe friction winder was installed on the No.1 shaft.

New winding arrangements were installed at the redesigned No.2 shaft pit-bottom in September 1962, together with 8ton-capacity skips to raise coal from the Parkgate and Thorncliffe Seams. In total the reconstruction schemes completed in 1962 cost some £4million.

In June 1961 a rapid loading 'merry-go-round' system, one of the first in the country, was commissioned. Comprising a 2,300 metric ton-capacity, pulverized coal smalls, concrete bunker with a travelling discharge chute, the system was capable of loading a 1,000 metric ton train, with a turn-round of sixty minutes.

Between November 1967 and October 1969, a 1,550yd-long drift, at a gradient of 1 in 6.7, was driven at the east end of the pit yard at Hickleton Colliery, to meet the Goldthorpe/Highgate workings. A pair of new fans were installed at Hickleton and commissioned in January 1970, the drift serving as a common ventilation return.

By the late 1960s things were not looking good for Hickleton Main. In March 1968 the *Colliery Guardian* reported, '**Hickleton Main in trouble**', the Doncaster Area Director was reported saying, '...Hickleton Main, South Yorkshire, which only two years ago was one of Britain's top fifty pits, is now under threat of closure in six months unless profitability is stepped up'. He went on to describe the absentee rate at the pit as 'Fantastic – the worst in

Hickleton Main Colliery, December 1978.

Britain, and probably the world'. The NCB had written off £600,000 in the first eleven months of that financial year (1967/68) which amounted to a loss of 19s 9d (almost £1) on every ton of coal produced. The men, it was said, had lost their confidence due to geological conditions and NCB planning decisions that had gone wrong.

Production from the Newhill and Meltonfield Seams ceased in 1983. Following a period of limited production, to allow extensive underground development to take place, all future output was obtained from the lower Parkgate, Thorncliffe and Silkstone Seams.

In the 1980s Hickleton was producing coal for the domestic, industrial, manufactured fuel (coke) and electricity generation markets.

In March 1985 the colliery was put on a development only basis. It had been the intention to bring Hickleton into production once the reserves in the Goldthorpe side of the combined mine were used up, but with the transfer of reserves in the Shafton Seam from Cadeby Colliery in 1986, Goldthorpe was able to continue at its level of production for an extended period of time. The additional investment needed to complete the Hickleton development could not be justified, and by late-1987 the colliery was put on a 'care and maintenance' basis.

Adverse geological conditions led to the abandonment of workings in the Thorncliffe Seam, and the NUM strike of 1984-85 struck the death-knell for the colliery. The colliery strategy was then reviewed resulting in the stoppage of the development of the Silkstone Seam, and continuation of the drivage to access reserves in the Parkgate Seam.

On 1 January 1986 Goldthorpe and Hickleton Collieries were merged for administrative purposes, the unit becoming known as Goldthorpe/Hickleton Colliery. Hickleton was kept open on a 'care and maintenance' basis, and operated on a 'development only' basis, to enter new reserves in the Parkgate and Thorncliffe Seams. Hickleton finally closed in 1988, the remaining men transferred to Goldthorpe on 31 March 1988. In 1990-94 the shafts were filled and capped.

Seams Worked:

Newhill Seam, 360yds, 5ft 3in thick
Meltonfield Seam, 385yds, 4ft thick
Barnsley Bed Seam, 538yds, 5½-8ft thick. This seam furnished a good quality house and coking coal, and the Hards a superior steam-producing quality coal.
Dunsil Seam, 554yds, 3ft 4in thick
Parkgate Seam, 789yds, 5ft 4in
Thorncliffe Seam, 827yds, best seam section 6ft 7in.

Highgate Colliery

Situated at the west end of the village of Goldthorpe, the colliery was close to the main Doncaster to Barnsley road, and mid-way between the two towns. The colliery was served by the Dearne Valley line of the LNER.

Highgate Colliery began in 1916, by the Highgate Colliery Co., who retained ownership until nationalization, with the sinking of two shafts to the Shafton Seam, at a depth of 50yds. The shafts were sunk with the objective of working some 600 acres of the Shafton Seam located to the north and west of the shaft pillar. The No.1, downcast and coal winding shaft was 10ft in diameter, and the No.2, upcast and man winding shaft 8ft in diameter. In addition, a drift was driven from the West Plane in the Shafton Seam to the surface, which surfaced 50yds north of the No.1 shaft.

Like its sister pit at Goldthorpe, because of the shallow depth of the workings, Highgate was designed to produce 1,000tons per week from hand got pillar and stall workings. In later years electrically driven Siskol percussion machines were employed at the colliery. A report in the *Iron & Coal Trades Review* in 1936 described the use of electrically driven Siskol coal-cutters in the Shafton Seam. The cutters worked the coal in panels, two machines per panel, one shearing the coal and the other undercutting it. The shearing reduced the amount of explosive used, which resulted in a larger percentage of round coal. The Shafton yielded a large amount of water, 400gal. per minute in the 1940s, which made the workings very wet and required continuous pumping.

Output and productivity were increased in the mid-1940s, with the introduction of American shortwall machines and scraper-chain conveyors.

In the early 1940s the colliery had screening plant capable of handling fifty tons per hour, but no washery. Neither was there any air compressing plant.

In the early 1940s and 1950s additional areas of coal were acquired, adding substantially to the colliery's reserves, as follows:

1943, 385 acres, ex-Doncaster Amalgamated Collieries.

1947, 1,000 acres, resulting from Nationalization.

1951, 850 acres, ex-NCB North-East Division No.4 Carlton Area.

At the close of 1952 reserves in the Shafton Seam were estimated as sufficient to last for forty years at an extraction rate of 1,000tons per day.

In 1946 the first of two drifts was driven from the surface to the northern part of the take. The second drift was started in the early 1950s and completed soon after. The first face was opened up from the bottom of the drift in 1956. Following completion of the drifts, the shafts were filled in September 1959, and sealed in April 1960, coal winding having ceased in June 1952, when all the coal in the original leasehold area had been exhausted. The intake drift was used for transporting coal until the colliery was linked with Goldthorpe Colliery in 1966, and the return drift was used for transporting men and materials until the colliery closed in 1985.

Although connected by rail with the Dearne Valley section of the London Midland Region railway, by the early 1950s almost half of the colliery's output was distributed by road. Power for the colliery was supplied solely by the area ring main, this was in operation in the early 1950s.

The first longwall faces were tried in 1953. These were unsuccessful due to poor roof conditions and associated water problems. Further trials took place in 1956, but these also failed for the same reasons. During this time further development had taken place in the Northern Area, where the workings further to the dip had an increased cover (ie. depth) of about 130ft. This increased depth resulted in more favourable conditions and longwall working proved to be successful, with some faces producing record outputs up to 16.54tons output per man-shift per face.

In 1959-60, 140 miners from Highgate were transferred to other pits because of a drop in demand for the type of coal being produced at the colliery. Following a sudden revival in demand for Shafton coal the NCB took the decision to mechanize the colliery. The following article appeared in the *Yorkshire Evening Post* on 16 June 1961:

> ***FAMILY PITS' SMASH THE OUTPUT RECORDS***
> *Doncaster's two smallest pits – Highgate and Goldthorpe, the 'family pits' where fathers, sons, brothers, uncles and nephews work together – are smashing coal production records...They are now one of the few mines in the country producing more than three tons per man on each shift. Their best average weekly total is...more than double the national average figure...Almost the whole of the output from the twins is fed to power stations, such as those at Doncaster and Mexborough...Two problems which make the performance of the 'identical twins' all the more remarkable is that both have water problems and both have to to use their machines on a shorter face than most pits - only 140yd long.*

Face mechanization, with shearer-loaders loading the coal onto armoured face-conveyors, was installed in 1962.

In 1966 Highgate and nearby Goldthorpe Colliery were combined when an underground link was made between the two units. In the following year a Colliery General Manager was appointed to oversee the combined mine. Eventually, all output from the combined mine was transported to the surface at Goldthorpe, for processing. Production in the Highgate area of the combined mine ceased in 1985, following the year-long NUM strike.

Seams Worked:

Shafton Seam, 5ft 3in thick over the area worked, of which the top 4ft 6in was worked.

Monckton Collieries

Located some seven miles north-east of Barnsley, and one mile east of Royston railway station, the colliery was located on the main line of the Midland Railway. The Monckton Main Coal Co. Ltd was registered on 16 September 1874. The Memorandum and Articles of Association for the new company provided for a capital of £120,000, divided into 1,200 shares of £100 each. From the outset Monckton Colliery was a large and ambitious project.

The colliery was well-served by local transport, adjoining the Barnsley Canal (opened 1799) where boat-loading staithes were built following an agreement with the Aire & Calder Navigation Canal Co. The colliery was also connected by a mile long branch line to the main line of the Midland Railway. By 1885 the colliery was also connected to the Manchester, Sheffield & Lincolnshire Railway, with sidings and a junction at Old Royston.

The first sod for the new enterprise was cut by Lord Galway, the Chairman of the Monckton Coal Co., on 24 May 1875. Lord Galway died at the end of the year, to be succeeded by his son, the 7th Viscount, who was to remain chairman of the company for the next fifty-five years. Two 16ft diameter shafts, located 75yds apart, the No.1 and No.2 pits, were sunk to the Barnsley Seam, at a depth of 475yds.

A clay mill, for manufacturing bricks, was erected in June 1875 and shaft sinking commenced in July. Mr James Beaumont contracted to sink the two shafts at a price of £15 5s per lineal yard. At the close of 1875 the No.1 shaft had reached a depth of 71yds and No.2 a depth of 65yds. Sinking proceeded during 1876, and by December, No.1 shaft had reached a depth of 254yds and No.2 shaft 286yds. In September 1877 the Barnsley Seam was reached.

By February 1878 No.1 shaft had reached 479yds and No.2 shaft 493yds. In March 1878 No.2 pit was producing sixty tons of coal per day, from the Barnsley Seam, which rapidly increased to 1,000tons per week by June and 2,500tons per week by December of the same year. All the seams under the property were understood to be unworked, with the exception of the Shafton Seam, which was being worked from a sloping drift by the Hodroyd Coal Co.

A range of workshops at the colliery included a truck shop for truck repairs, a tub repair shop, a fitting shop, a joiner's shop and a smithy.

During 1878 an attempt at manufacturing coke using unwashed slack was tried, but the results proved unsuitable for steel-making.

In February 1879 work commenced on the erection of twelve beehive type coke ovens, which went into production in October 1879. By February 1880 twenty-four coking ovens were in operation, and the output from the No.2 pit had increased to 650tons per day. Further coke ovens were added during 1880, so that by 1881, a total of forty-two ovens were in operation. In 1884 ten Jameson's Patent By-product Recovery coke ovens, producing by-product oil, were erected. The number of coke ovens at the colliery increased steadily, with 135 in operation by late 1889, and 153 by 1890.

A gas plant manufactured by Messrs Newton & Chambers was brought into production in 1886, providing gas to light the whole of the surface.

New Monckton Collieries

In 1901 a reconstruction of the undertaking took place, the old Monckton Coal Co. Ltd was wound-up and a new company, with a transferred capital of £220,000, was formed and called New Monckton Collieries Ltd. New Monckton Collieries, was incorporated on 25 July, and took over the business and assets of the company as a going concern. These assets comprised the Monckton Main and Hodroyd Collieries, 180 beehive coke ovens, by-product plant and a brick yard. New Monckton Collieries Ltd was floated as a public company, and the first shares to be allocated to the public were approved at a meeting in August 1901.

At the start of the twentieth century the colliery had an annual output of 470,000tons, the coking plant had a capability of 1,800tons per week, and the brick works a capacity of four million bricks per annum. New technology, in the form of electricity and compressed air, was employed to power haulages, pumps and coal-cutting machines by 1902, and portable, electric safety-lamps were in use by 1912.

By December 1901 drifts, at a gradient of 1 in 3, were being driven to work the Haigh Moor Seam, which lay below the Barnsley Seam. In 1913 two further drifts were sunk from the Barnsley level to the Haigh Moor Seam. In 1912 the Meltonfield Seam was being developed at the colliery.

In 1920 the coke ovens and by-product plant at the colliery were transferred to the Monckton Coke & Chemical Co. Ltd, an agreement having been entered into to ensure interchange of services between the two concerns.

The first steps towards modernization of the colliery began in 1922 with the erection of concrete coal storage bunkers. In 1924 the original No.1 battery of thirty-six waste-heat ovens were replaced by a new battery of forty-five waste-heat ovens, and in 1927 the No.2 battery of waste-heat ovens was replaced by a battery of twenty Simon Carves silica-underjet regenerative ovens.

By April 1924, the No.1 and No.2 pits had been working for almost half a century, and had workings extending for a radius of two miles. They were reaching their limit of profitability. In order to maintain the production of the outer reaches of the Haigh Moor Seam a number of options were considered. It was decided to sink a new shaft, No.5, at the western extremity of the coalfield for man-riding, raising coal and to improve ventilation. In November 1924 the board passed the resolution to proceed with a sinking near Notton Grange Farm. Preparatory work proceeded, but due to the slack trade in the industry, and the General Strike of 1926, sinking was delayed. Sinking of the No.5 shaft started on Monday 11 July 1927, the work contracted to the Francois Cementation Co. of Doncaster. The shaft reached the Barnsley Seam, at a depth of 317yds 2ft, on 31 August 1928, and reached the final depth of 326yds on 8 September. By November some 700 men were using the shaft.

In 1938 headings were driven into the Kent's Thick Seam, at a depth of 406yds, from the No.1 shaft. In the same year the go ahead was given for the construction of ambulance rooms and pithead baths at the No.1 and No.2 shafts and the newer No.5 pit. In early 1941 the baths at the No.1 and No.2 shafts came into operation.

In the period 1938-1939 the No.3 and No.4 batteries of waste-heat coke ovens were replaced by the first battery of Becker regenerative ovens. These were erected by the Woodall-Duckham Construction Co.

By the early 1940s the Barnsley and Haigh Moor Seams were reaching the limits of their working boundary in the No.2 pit, and other seams were considered for exploitation. The

New Monckton Nos 1 & 2 Colliery, early 1900s.

Winter Seam, at a depth of 348yds, was decided upon, but not carried through at the time. The worsening financial situation at this time led to the proposal to close the No.2 pit, with the exception of development work in the Winter Seam, re-deploying the men in an effort to minimize costs. The outcome was to keep the pit working. On 4 February 1946 two new faces in the Winter Seam came on stream.

In the early 1940s various schemes were proposed to concentrate the workings at the colliery and reduce the number of employees, by re-deploying them to other collieries. In February 1944 the government's Production Grouping Scheme tied Monckton Collieries to the nearby Fountain & Burnley Group of collieries.

In the run up to nationalization, the company was called upon to provide information as to the value of their assets, in order to determine the value of compensation to be paid. The company arrived at a figure of £1.7million.

In the period 1946-1947 a second battery of Becker Regenerative ovens was built, followed by a third battery of eleven ovens in 1952-1953. The first battery of ovens was replaced in 1953-1955.

Following nationalization a reconstruction scheme was announced, costing an estimated £7.4million. The scheme included electrification of the colliery, sinking a new (No.6) shaft, constructing new headgear at No.1 shaft, deepening No.3 shaft, installing a new fan at No.4 shaft, and construction of new coal-preparation plant, offices, workshops and lamproom.

At Vesting Day, the Barnsley and Haigh Moor Seams were almost worked out and the output of the colliery was some 13,000tons per week. The surface plant was over fifty years old and in poor condition, due to lack of investment over the years. The shafts were either too narrow or positioned in such a way that the total output could not be brought to one shaft. It was decided that a new, large diameter, shaft would be sunk, which would be capable of raising up to 7,000tons per day of coal, and the surface would be totally modernised. Coal from Monckton Collieries No.3 and No.4 would be brought to the new No.6 shaft via a locomotive tunnel, and raised at the new shaft. By 1959 some seventy-five per cent of the reconstruction was complete, and the annual output of the colliery had increased from 695,119tons in 1947, to 812,146tons in 1957. The final annual output was expected to be 1.25m tons.

The combined New Monckton Collieries worked a coalfield of some 5,000 acres, with extensive reserves of high-quality coking coals, regarded as the most valuable in Europe, estimated at 150m tons. These figures more than justified the reconstruction scheme at the colliery.

The reconstruction scheme started in 1953 and took until 1959 to complete. The scheme was designed to concentrate output at a number of underground loading points, from where the coal was hauled to the shaft bottom by locomotive haulage. At the surface it was decided to establish new sidings, electrify the winders, and build a new washery. The scheme was approved in 1953 and work proceeded in October of that year. The No.1 pit heapstead, which was in an unsafe condition, was rebuilt. A 600ton per hour coal preparation plant, by Simon Carves, was erected and commissioned. The 24ft diameter No.6 shaft was sunk to a depth of 820yds, on the No.1 and No.2 pit site, and skip pockets were constructed at the Melton Field and Barnsley Seam levels. The concrete Koepe winding tower over the No.6 shaft still stands today as a significant landmark in the area. Improvements in underground transport, brought about by the reconstruction of the 1950s, resulted in the outlying No.5 shaft becoming redundant. This shaft was decommissioned during the reconstruction. The effect of the reconstruction was to boost output from 13,000tons per week in 1947, to 15,760tons per week in 1959.

The investment pumped into New Monckton Nos 1, 2 and 6 by the NCB in the 1950s, resulted in a labour force of 1,930 mining the Lidgett, Kent, Beamshaw and Barnsley Top Seams.

New Monckton Collieries comprised two units, New Monckton Nos 1, 2 and 6 and New Monckton Nos 3 and 4. The combined take comprised an area of 5,000 acres. The colliery take was separated by a wide strip of barren ground, which was the reason why the company established the two separate collieries. Coal was wound from shafts 1, 2 and 3. The coal from New Monckton shafts 3 and 4 was transported overland by endless-rope haulage to the washery at New Monckton 1, 2 and 6.

The NCB had run the two production units in tandem until closure on 3 December 1966. For the first time the reasons for the closure of a colliery, were given as economic rather than on grounds of exhaustion. After the initial vigorous opposition to closure from the unions, the recommendations were accepted. At closure New Monckton 1, 2 and 6 employed 1,273 men and Nos 3 and 4 employed 1,000 men. Following closure the majority of the workforce were relocated to neighbouring collieries.

Seams Worked:

Meltonfield Seam, 313yds deep, 3ft thick.
Beamshaw Seam
Kent Seam, 406yds deep, 2ft 9in thick.
Barnsley Top Seam, 4ft thick, 470yds deep.
Barnsley Seam, 475yds deep, 5ft 3in thick.
Haigh Moor Seam, 651yds deep, 4ft 5in thick.

Rotherham Main Colliery

Rotherham Main Colliery was located about a mile south of Rotherham railway station, on the Midland Railway, and close to the Tinsley-Bawtry road. The owners of the colliery were John Brown & Co., of the Atlas Steel & Iron Works, Sheffield, who also owned the nearby Aldwarke Main and Carr House Collieries, and retained ownership for the colliery until nationalization. The colliery was served by the LMS.

The Company commenced the sinking of the colliery in 1890. Two, 18ft diameter, shafts were sunk to the Parkgate Seam, at 602yds, by 1893, passing through the High Hazels Seam at 258yds, the Barnsley Seam at 351yds and the Swallow Wood Seam at 434yds. The No.1, or North, downcast shaft and the No.2, or South, upcast shaft were located 100ft apart in a north-south line. During shaft sinking heavy feeders of water were given out from the Rotherham Red Rock, a fissured sandstone which was passed through. In order to dam the water back, 69yds of cast iron tubbing were inserted into the No.1 shaft, and 86yds into the No.2 shaft. Beneath the tubbing there was no water to contend with and the shafts were lined with ordinary red bricks and mortar.

Two well-known faults, the Spa and Holmes, as well as a number of smaller faults, crossed the coalfield in a north-westerly to south-easterly direction. These formed a large trough, which necessitated the driving of drifts from the pit-bottoms to workable areas of coal. Drifts were driven, through the faults, from the Parkgate level to the Barnsley Seam above, and the Silkstone Seam at

Rotherham Main Colliery, early 1900s.

a lower level. The winding-engine at the north or downcast shaft wound Parkgate coal exclusively.

At the upcast shaft there were two engines placed at right angles, the south engine drawing coal from the Barnsley Seam and the east engine from the High Hazels Seam in the same shaft. An auxiliary fan, placed 2,000yds from the shaft bottom, was arranged to ventilate a district cut off from the pit by faults. This was a 45in diameter Sirocco fan, driven by a geared belt drive from a compressed air engine.

Coal processing plant comprising a standard Baum type washer, capable of handling seventy-five tons per hour, and sixty-five Simon Carves type coke ovens, in two batteries, a tar distillation plant, and a by-product plant producing sulphate of ammonia, benzol and tar.

In 1907 haulage drifts were driven from the Parkgate to the Silkstone Seam. These drifts were some 400yd-long at a gradient of 1 in 4.

In 1915 the colliery was working four seams, the High Hazel, Parkgate, Barnsley and Silkstone. Power at the colliery was derived principally from exhaust steam turbines, and was used for rock drilling, pumping, hauling and ventilation.

In November 1926 arrangements were made to pump from the High Hazels shaft at Holmes Colliery, in order to work an area of Barnsley coal to the north-west of Rotherham Main. Pumping did not commence until May 1929.

In the early 1940s some forty per cent of the colliery's output went to the coke oven and by-product plant attached to the colliery. The twenty-eight coke ovens were carbonizing 3,000tons of coal a week. At this time the colliery was working the Barnsley, Parkgate and Silkstone Seams. The coal wound from the working seams was raised at the downcast shaft, from the Parkgate level. Coal from the Silkstone Seam was brought up a rising drift from the seam, and that from the Barnsley Seam via a dipping cross measures drift.

The Mines Department Technical Adviser report, published in March 1942, reported that the colliery output was very low and the remaining reserves could be easily worked from Aldwarke or Silverwood Collieries, and that there was no reason why the pit should not close. Workings in the Barnsley and Parkgate Seams were over two miles from the pit-bottom, the Parkgate Seam suffered with a poor roof, and the Silkstone Seam was estimated to have three years left to run. The colliery closed in May 1954.

Seams Worked:

Seam	**Depth (yds)**	**Thickness (inches)**
High Hazels Seam	258	42
Barnsley Seam	351	61
Swallow Wood Seam	434	71
Parkgate Seam	608	60
Silkstone Seam	688	53.5

South Kirkby Colliery

The colliery was situated midway between Doncaster and Wakefield, and at an equal distance from Barnsley. It was sunk by the South Ferryhill & Rosedale Iron Co. of County Durham, who had leased a large area of land in the village of South Kirkby. The colliery was the first in the district to be sunk to the Barnsley Seam. Sinking of two, 15ft-diameter shafts, located 150yds apart, started in 1876 and by August 1878 the Barnsley Seam had been reached at a depth of 634yds. Both shafts were lined with cast iron tubbing, due to the water released during sinking.

The No.1 downcast shaft was sunk to the Barnsley Seam, and the No.2 upcast shaft to the Haigh Moor Seam, at a depth of 712yds. In 1897 a 14ft diameter staple-pit was sunk from the Beamshaw Seam, at a depth of 509yds, to the Haigh Moor Seam, passing through the Barnsley Seam. In 1897 a staple shaft was sunk from the Beamshaw Seam to the Haigh Moor Seam, enabling all three seams to be worked.

By 1879 the colliery had cost some £76,800 and incurred liabilities of £230,000, sums which were more than had been anticipated, and the colliery was yet to provide a return on this investment. By late 1879 the project stopped and remained so until the following spring. During this time the colliery was unsuccessfully offered for sale. It looked as though the new venture was to fail, when the South Kirkby, Featherstone & Hemsworth Collieries Co. was formed, with a capital of £100,000, to take over the venture. Following the take-over, work to develop the colliery proceeded rapidly, with the opening-out of the Barnsley and Haigh Moor Seams, and construction of surface plant.

Until 1921, when an electricity sub-station was built, the colliery relied entirely on steam and compressed air power. From 1921 the requirement for electric power grew and in 1924 a power station was built at the colliery. A supply was taken from the power station to the colliery company's Hemsworth Brick Co., and by 1934 all collieries, anciliary works and pumping stations in the group were connected to the power station.

By the mid-1920s the entire output was produced from the Barnsley and Haigh Moor Seams, the Shafton and Beamshaw Seams having been temporarily stopped, awaiting more prosperous days. By 1939 the Haigh Moor Seam was exhausted and ceased production, and by the early 1950s the Barnsley Seam was nearing exhaustion.

At nationalization the principal issue for the colliery was the small diameter of the shafts. Because of the large reserves of coking coal it was decided to sink a new 24ft diameter shaft. By 1959 the shaft was some 300yds deep.

In the early 1950s a £2 million reconstruction scheme was implemented. The scheme had the objective of increasing annual output to 690,700tons. It entailed the working of two new seams, the Meltonfield and Winter, the installation of skip-winding, electrification, and the construction of new coal preparation plant and railway sidings at the surface. The reconstruction was planned for completion by 1957-58. The developments were planned to give access to 55m tons of reserves, which it was anticipated would give the colliery a life of a further fifty-five years. Manpower was expected to rise to over 2,000. On 28 January 1957 the *Yorkshire Post* reported: 'South Kirkby Colliery's output last week was 13,438 tons, and was the pit's best since Vesting Day, being 23 tons better than the last record set up in November, 1955'.

In July 1967 South Kirkby and Hemsworth Collieries, which had been physically linked underground for many years, were reorganized into a single unit, known as South Kirkby Colliery, with a manpower of 2,718 men. All the coal produced at Hemsworth was wound at South Kirkby, where two of the three shafts were fitted with balanced skips. The shafts at Hemsworth were retained for ventilation. However, things were not going well, and on 22 September 1967 the *Colliery Guardian* reported, '**South Kirkby Shock**'.

The colliery, which was on the list of fifty long-life pits drawn up by the NCB, was losing money heavily and its future was at risk unless the situation improved. The recent reconstruction and merger with Hemsworth Colliery had cost £5million and the colliery had lost £1.3million in the previous year.

South Kirkby Colliery.

South Kirkby Colliery, No.1 (foreground) and No.3 pits, in the early 1980s.

In June 1969 production started from Riddings Drift, a new drift with its surface entrance sunk in the pit yard at South Kirkby Colliery. The colliery was ventilated by the upcast at South Kirkby and shared the same surface facilities, but was operated as a completely separate unit.

In the 1970s the colliery was working the Newhill, Meltonfield, Beamshaw, Dunsil and Barnsley Seams. In 1979 the coal preparation plant at South Kirkby was treating coal produced by South Kirkby, Kinsley Drift and Ferrymoor/Riddings Drift.

In 1974 reconstruction plans were initiated, in the NCB Barnsley Area, to increase output and overall efficiency. A major complex was created at South Kirkby to receive the output from Ferrymoor/Riddings Drift, Kinsley Drift and South Kirkby, at a central coal preparation plant.

In March 1987 South Kirkby Colliery merged with Ferrymoor/Riddings and, having incurred heavy losses, was under review. On 25 March 1988 South Kirkby Colliery closed.

Seams Worked:

Shafton Seam, 185yds deep, 4ft 11in thick seam section.
Meltonfield Seam, 448yds deep, 3ft 6in thick.
Winter Seam, 489yds deep, 2ft 4in thick.
Beamshaw Seam, 509yds deep, 2ft 11in thick.
Barnsley Seam, 634yds deep, 4ft 9in thick.
Dunsil Seam, 642yds deep, 3ft 3in thick.
Haigh Moor Seam, 712yds deep, 3ft 9in of the seam worked.
Fenton Seam

9
Mining 1901-1910

The 1900s saw the first large-scale development of the concealed coalfield of South Yorkshire. Coal mining had shifted towards Doncaster, with the sinking of the first of the Doncaster area collieries at Brodsworth and Bentley. Brodsworth Main was soon to become the largest colliery in Yorkshire, and following the sinking of the third shaft, the highest output three-shaft colliery in Britain.

Other large collieries sunk in this period include Silverwood and Frickley, which pierced the Maltby and Frickley Troughs. Both collieries were working the Barnsley Seam at depths far in excess of that worked by their neighbours.

This was the start of the development of the Doncaster area of the coalfield, and the era of the large colliery designed for outputs of a million tons. These collieries were working the Barnsley Bed Seam at depths of 600yds and more and frequently experienced problems in shaft sinking due to the water bearing sandstones which were encountered. Because of the large initial capital investment required, and the high development and running costs, these collieries had to be worked on a large scale. Typically they were working mineral royalties of 6,000 acres.

The Doncaster area of the South Yorkshire Coalfield attracted the attention of mining companies from Derbyshire. They were concerned about the gradual exhaustion of the Top Hard Seam (the Derbyshire and Nottinghamshire Coalfield equivalent to the Barnsley Seam) and were willing to go further afield in search of it. This relentless eastwards drift of mining would ultimately lead to Doncaster becoming the centre of the South Yorkshire Coalfield, as Barnsley had been before it.

The magnitude of the sinkings in the Doncaster region would lead to the formation of a number of joint ventures to share the expense and associated risks of developing collieries in this part of the coalfield. Brodsworth and Dinnington were two such collieries, sunk as joint ventures, which would be followed in later years by Rossington and Upton Collieries.

By the beginning of the twentieth century, the Barnsley Seam, in the central region of the coalfield, was nearing exhaustion and collieries were being sunk to work the deeper Parkgate, Haigh Moor and Silkstone Seams. As well as the deepening of existing collieries in the central region of the coalfield, collieries were also being sunk to work the shallow seams such as the Shafton. To this end collieries were sunk at Goldthorpe, Brierly and South Elmsall

Collieries sunk in this phase include:

Barnburgh Main – sunk 1915 as extension to Manvers Main
Bentley
Brierley
Brodsworth Main
Dinnington Main
Frickley
Goldthorpe
Silverwood

Barnburgh Main Colliery

Situated half a mile south-west of Barnborough village and six and a quarter miles west of Doncaster, Barnburgh Main Colliery was sunk to the Thorncliffe Seam between 1912-1915 by Manver Main Collieries Ltd, who retained ownership for the colliery until nationalization. Situated approximately three miles from Manvers Main, Barnburgh Main was connected to the processing plant at Manvers by a private railway line. In the 1930s the line carried 25,000-28,000tons of coal per month between the sites. Barnburgh was linked to the LMS Railway, and to both the LMS and LNER via Manvers, thus providing a number of alternative routes for disposal of the collieries' output. Barnburgh Colliery comprised shaft Nos 5 and 6 of Manvers Main Collieries, shaft Nos 1, 2, 3 and 4 being located at Manvers Colliery site. The 18ft diameter No.5 shaft was the downcast and the 16ft diameter No.6 shaft the upcast at the colliery. As well as the two main working shafts, there was also the 12ft diameter, 86yd deep, No.7 well shaft, sunk to the Shafton Seam to provide water for the colliery. The Shafton Seam furnished excellent quality water, which did not require softening.

The No.5 and No.6 shafts were sunk almost simultaneously, sinking starting in June 1912. The Barnsley Seam was reached in the No.6 shaft in May 1914, at 508yds, and in the No.5 shaft in June 1914. The Parkgate (or Deep) Seam was reached in February 1915, at a depth of 757yds. Both shafts were brick lined throughout, with no tubbing and were equipped for coal winding. Sinking of the No.7 pumping shaft commenced in December 1912 and reached the Shafton Seam in April 1913. The Barnsley Seam provided the total output of the colliery until 1938. The Barnsley Seam had a working section 6ft 5in thick, leaving 2ft 5in of top coal unworked.

The colliery was well-equipped with workshops, compromising fitting, blacksmiths and electricians shops, a joiners shop and saw mill. All the workshops were contained in one block of buildings. Other buildings on the site included; the offices, ambulance room, stables, and garage. The majority of the surface plant was designed and erected by Qualter, Hall & Co. of Barnsley. Spiral separators, erected in April 1924, were the first to be commissioned in this country. The power plant at the colliery was connected to Manvers Colliery and the 10,000V Doncaster Collieries Association ring main. Electric lighting, for the nearby villages of Barnborough, Adwicke and Harlington, was provided by the DC generator at the colliery. In 1932 a handsome pithead baths, 202ft long by 85ft wide, was erected entirely at the colliery company's expense. Some 850 men at the colliery subscribed at a rate of 6d per week.

In 1937 the Parkgate Seam was opened out, in readiness to take over from the Barnsley Seam, which was becoming rapidly exhausted. At this time the Barnsley Bed Seam was producing 2,300tons per day, wound in two shifts in the No.5 downcast shaft. The Parkgate Seam output was 2,000tons per day, and in the interests of concentration it was desirable to wind the total output in one shift in the No.6 shaft. Calculation showed that it was impossible to wind the total output of 2,000tons in a single seven-hour shift using tubs, and the only option was to employ skip-winding. Installation of skip-winding, using 8ton skips, took three months and on 30 August 1937 coal was being wound by skip in the No.6 shaft. By 1944 over 2.5m tons had been wound by skip and the facility was working well. Up to that time the record output had been 2,119tons in seven-hours, which equated to 285 skip loads in seven-hours. This was the first permanent installation of skip-winding in the Yorkshire Coalfield, and one of the first in the country. Barnsley coal continued to be drawn in the No.5 shaft at this time.

Electric and compressed air powered haulage was used extensively in the Barnsley Seam. Main and tail haulage in use on the main levels and single rope haulage in the gate roads, which were on the dip. By 1930 there were only three pit ponies at work in the colliery.

In the early 1940s the No.6 upcast shaft was drawing Parkgate and Thorncliffe coal from the Parkgate level, the Thorncliffe coal transported up a pair of drifts driven from the Thorncliffe level. The No.5 downcast shaft was drawing Newhill coal from the Newhill level, Barnsley and Haigh Moor coal from the Barnsley level, the Haigh Moor coal transported up a drift from the Haigh Moor level. The Newhill Seam was 5ft 3in thick and the full seam section was being worked. The Barnsley Seam was 11ft 3in thick, of which 5ft 10in was being worked. The full 3ft 10in, including a 3in thick dirt band, of the Haigh Moor Seam, and the full 5ft of the Parkgate Seam, were also being worked. The Newhill Seam was the other seam being worked and only the top 2ft 10in was being taken.

By the early 1940s the two mainstays of the colliery, the Barnsley and Parkgate Seams were facing difficulties. The Barnsley Seam was almost worked out. The Parkgate Seam was suffering from a large washout running north-south, about 750yds from the pit-bottom. Replacement capacity was required urgently. Developments were speeded up in the Haigh Moor, which was just being developed, and the Newhill and Thorncliffe Seams. A trial roadway was driven 138yds across the washout in the Parkgate Seam and coal was found again. The Mines Department Technical Advisor's report of March 1942 stated: 'Due to the short life left in the Barnsley and the uncertainties in the Parkgate Seam with large areas washed out, a considerable amount of long term development and exploration has been necessary.'

At this time, a second drift was being driven between the Barnsley and Haigh Moor Seams. There was no washery at Barnburgh and all coal for washing was taken three miles to the Manvers site by a single-track private railway.

From 1942 the Swallow Wood Seam was being worked, but was found to be of inferior quality and was soon abandoned. At Vesting Day four seams; the Newhill; Barnsley; Swallow Wood and

Barnburgh Main Colliery, October 1989.

Parkgate, were being worked. The Newhill Seam being developed to replace the Swallow Wood.

A major reconstruction scheme took place in the period 1950-56. It included: driving a horizon at the Winter Seam level; installing locomotive haulage and six ton mine-cars; sinking two staple shafts between the Newhill and Winter Seams; deepening the No.5 shaft to the Thorncliffe Seam, at a depth of 788yds, and constructing an overland rail link to the nearby coal preparation plant at Manvers. From 1958 all Barnburgh Main coal was transported via the private branch line to Manvers. Reconstruction of the pit-bottom enabled all of the output to be wound in one shaft, and enabled the colliery to become part of the £6.5million Manvers Central Scheme. In the late 1940s coal reserves at the colliery had been estimated at 107,750,000tons.

In 1950 a new electric winder, probably the largest in the country, was installed on the No.6 shaft at Barnburgh. Concentration of output from the colliery at one shaft necessitated the installation of a clutch-drum winder, to enable the colliery to wind from different levels. The 3,500hp winder was one of three manufactured in the USA, in 1944, for Russia. The new winder wound from 750yds, using the original 8ton-capacity skips. Ventilation at the colliery was reversed, the No.5 shaft became the upcast and the No.6 shaft the downcast. Coal winding, using the new winder, was concentrated on the No.6 shaft and the No.5 shaft was used for man-riding and the transport of materials. The first skips of coal were wound with the new winder on 5 August 1950.

By the mid-1950s coal was being skip wound on two shifts per day, coal wound from the Parkgate Seam level, at 757yds, on one shift, and from the Winter Seam level, at 360yds, on the other shift.

The Meltonfield Seam was in production by early 1962 and by 1965 all of the output came from the Newhill and Meltonfield Seams. 1974 saw the start of retreat mining and by 1978 the colliery became the first in the South Yorkshire area to produce its total output by retreat mining. From June 1983, until production ceased on 16 June 1989, production was based solely on the Meltonfield Seam. In the latter years Meltonfield coal was washed and used to 'sweeten' other coals used for power station fuel. Attempts to develop a new area of the Newhill Seam in March 1988 proved unsuccessful, due to adverse conditions, and the decision to close the colliery was taken in February 1989. On 6 May 1989 the *Daily Telegraph* reported:

Loss-making pit employing 750 to close next month.
Barnburgh Colliery in South Yorkshire is to close next month. Its 750 workers will be offered either redundancy or jobs in other collieries...Last year Barnburgh lost £2.8 million.

On 16 June 1989 Barnburgh Main Colliery ceased coal production and closed. Geological difficulties and a lack of reserves had closed the colliery. Following closure, the No.6 headgear was blown up on 8 June 1990, and the No.5 headgear on 10 August 1990.

Successful exploratory borings near Barnburgh resulted in provisional plans for a drift mine in the area, similar in concept to Riddings Drift. A drift mine would take two years to develop, compared to ten years for a deep mine. In the early 1970s the NCB had plans to operate a drift mine from the pit yard at Barnburgh. The plan was to produce an annual output of 500,000tons from 4,000 acres of the Shafton and Cudworth Seams. There were estimated to be 5m tons available in the Shafton Seam and 3m tons in the Cudworth Seam. It was estimated that the development would cost £750,000 and employ 300 men. Coal from the drift would be transported overland to the Central Coal Preparation plant at Manvers. It is likely that, following the practice established at Riddings Drift, the Barnburgh drift would have shared facilities with Barnburgh Colliery, but would have been regarded as a separate unit. With the run-down of the industry the project never started.

Seams Worked:

Newhill Seam, 294yds deep.
Meltonfield Seam, 328yds deep, 48in thick.
Barnsley Seam, 508yds deep, yielded 6ft 6in of workable coal, and 2ft of inferior top coal or 'bags' which was usually left intact, except at the gates where extra height was required.
Swallow Wood (or Haigh Moor) Seam, 566yds deep.
Parkgate Seam, 758yds deep.
Thorncliffe Seam, 792yds deep.

Bentley Colliery

Situated some two and a half miles north of Doncaster, the colliery was sunk by Messrs Barber, Walker & Co., who retained ownership on the estate of Sir William Cooke, until nationalization.

Barber, Walker & Co. of Eastwood was a long established mining concern, tracing its origins back to c.1680 when the company was formed in Nottinghamshire. Like many of their mining contemporaries in the Nottinghamshire coalfield, the company was concerned about the gradual exhaustion of the Top Hard Seam (the equivalent of the Barnsley Seam in Nottinghamshire), and were willing to go further afield in search of it.

The colliery was connected to the main line of the Great Northern Railway, located about a mile to the east, and in 1914 a connection was made to the newly constructed Hull & Barnsley mineral line, located to the west. The royalty area of the colliery included 6,940 acres of the Barnsley Seam and 6,250 acres in other seams.

The core from a borehole at Daw Wood, sunk by the Vivian Boring Co. in 1893, close to the eventual location of the shafts, had shown 9ft of Barnsley Seam coal. At a special meeting of the company, held on 27 March 1902, Messrs Barber, Walker & Co. took up the option to work the coal and proceed with the sinking of a colliery in the centre of the royalty. In May 1903 steps were taken to appoint a manager to oversee the sinking and general layout of the pit. A manager, trained in the county, was appointed and took up his duties on 22 June 1903. Though he was deemed the most suitable, and came with excellent references from owners of collieries in the Wakefield area, under his direction progress had been slow and unsatisfactory.

Each visit of Mr T.P. Barber to the site increased his anxiety, and at a meeting of the company, held on 7 February 1905, he voiced his opinion that the manager was not competent to carry out the sinking. The manager was released and Mr J.W. Fryar, of Bettisfield Collieries, Bagillt, North Wales, was requested to go at once to Bentley, and report on the layout of the colliery and position generally. The company were impressed by Fryar and set out to acquire his services as full-time manager. However Fryar was committed at Bagillt, and could only commit to one day per week at Bentley. This was accepted as a temporary expedient, and Fryar took over responsibility for the sinking from 2 May 1905. Things went well and by the summer of 1905 the partners in the venture were convinced of the necessity of appointing

Fryar as a precursor to the success of the venture. The services of Fryar were secured and he commenced his duties as General Manager on 15 November 1905.

Sinking of the two 20ft diameter shafts at Bentley was watched with interest by mining engineers in South Yorkshire and further afield. The problem faced by the sinkers was the surface layer of alluvial quicksand which covered practically the whole area. Believed to be 50ft in thickness, the quicksand proved to be some 100ft thick.

Sinking of the No.2 upcast pit commenced on 9 October 1905. It was decided to sink through the quicksand bed by lowering bolted, cast iron tubbing and sheet piling, tongued and grooved into each other to form a complete circle, the piling being pushed down under pressure. Sinking was started on 9 October 1905, and the sinkers expected to reach the Triassic sandstone at a depth of 50ft, but as they met nothing but quicksand it was decided to put down a borehole in the shaft. The borehole showed that the Triassic sandstone was at a depth of 100ft, and not the expected 50ft. Problems during sinking had distorted and cracked the tubbing, so it was decided to abandon the sinking and re-sink a short distance away.

Sinking of the new shafts recommenced on 3 March 1906, with stronger tubbing and piles as well as other improvements to the original scheme. The new No.2 Pit shaft started at 23ft in diameter, with a finished diameter of 20ft. Further difficulties ensued but progress was made. At a depth of 50ft the flow of feed water increased to 100gallons per minute and gradually increased to 400gallons at a depth of 80ft. Sinking pumps were hung in the shafts to cope with the inflow of water. The top of the Triassic sandstone was reached at a depth of 100ft on 6 June 1906, after some eighteen and a half weeks. This achievement was perhaps the first of its kind in British coal mining, as no other colliery had experienced quicksand problems of such magnitude. Sinking of the No.2 shaft continued and the Barnsley Seam was reached in April 1908, at a depth of 625yds.

Experience gained from sinking the No.2 Pit made sinking the No.1 downcast pit easier. Sinking of the No.1 Pit started on 22 September 1906 and reached the Triassic sandstone, at a depth of 98ft, on 29 October, after only five weeks. From the base of the quicksand sinking of the shaft proceeded in the orthodox manner, reaching the Barnsley Seam without further problem in November 1908, at a depth of 624yds, and the Dunsil Seam at 642yds. A sump was formed at a depth of 654yds, some 12yds below the Dunsil Seam. The two shafts were finished with 20ft internal diameters and fitted to draw 2,000tons per day.

Construction of the surface buildings and works proceeded simultaneously with the sinking of the shafts. Records show that the total cost of the No.1 winding-engine, together with its housing, totalled £10,681. Similarly, the No.2 shaft cost £11,568. The boilers, fully equipped and laid down with pumps, well and induced draft, costing £6,774. The chimney cost £540, and water softening plant and reservoir £1,164. From the outset the colliery boasted a power station, equipped with a pair of Rateau turbines, built at a total cost of £13,859. In 1908 a screening plant was erected, described as second to none. The plant had a capacity of 3,000tons per eight hour shift. Well equipped workshops, comprising fitting, blacksmith's, spring, electrician and joiner's shops were provided. In addition there were saw and mortar mills, a timber yard and a storehouse. The covered-in heapstead, of reinforced ferro-concrete, was completed in 1911. It is believed that this was the first occasion that ferro-concrete was used for this purpose anywhere in the mining world. The building measured 176ft long by 34ft wide, and was 58ft from the foundation level to the top of the flat roof. The heapstead was built to the design of a Mr Mouchel. Ferro-concrete was used extensively at Bentley for the flooring of offices and stores, elevated gangways and stairways. The absence of exposed timber and steelwork reduced maintenance costs. A brick-built lamp cabin, measuring 53ft by 29ft, located close to the main entrance gate, was connected to the pit top by a sloping gangway. Fryar showed foresight in specifying a frequency of fifty cycles for the electrical installations at the colliery, when many new plants of the day operated at twenty-five cycles and were later faced with expensive conversion costs.

Heading out from the pit-bottom commenced during 1908. The labour force employed underground was selected from men employed in the same seam at the company's High Park and Watnall Collieries. The team proved to be loyal to the company and General Manager, and had a good influence on newcomers to the pit.

The surface of the land overlying the Bentley mineral area is predominantly flat and low-lying, some two thirds of it is less than 25ft above sea level. The area is drained by the River Don and had been subject to frequent flooding. Remedial drainage works had been carried out over the previous 300 years, resulting in a complex drainage system, comprising high and low-level drains. In about 1912 the effect of coal mining on surface drainage was first experienced, and remedial measures started. Records showed that subsidence, due to working the Barnsley Seam, varied from 3ft 9in to 4ft 6in. On 17 October 1914 the company entered an agreement with the Don Drainage Commissioners, which entitled them to enter and repair surface drains. However, problems ensued with Sir William Cooke, the landowner, which resulted in the company purchasing 1,006 acres of land from Cooke, and a further 396 acres from his Trustees, in order to secure unhindered control. Subsequently, further land was purchased, so that by the time of nationalization the company owned the freehold for about 2,392 acres of land.

When Mr Joseph Stafford became the Chief Engineer at Bentley Colliery, in 1921, with special instructions to tackle the drainage problem, he saw that the problem had to be viewed

Bentley Colliery, showing the reinforced ferro-concrete heapstead, June 1994.

on a wider scale than had previously been the case. Stafford set about devising a comprehensive drainage scheme, and though he went to Bentley with a superficial knowledge of the subject, he was soon to become a recognised expert in the field, being called upon by drainage authorities to give evidence and advice. It was said that Stafford's headquarters resembled a meteorological office. On 29 May 1929 the Doncaster Area Drainage Act was passed, followed by a further Act on 13 April 1930.

In September 1931, May 1932, March 1933 and February 1941 major floods occurred in the Bentley area. Caused by exceptionally heavy rainfall and melting snow in the Pennines, the River Don overflowed its banks and flooded thousands of acres of agricultural land. On each occasion livestock had to be moved and hundreds of families evacuated. In the floods of 1931 and 1933 the colliery was idle for a week, and for two weeks in 1932. The problem was the inadequacy of the water channel in the lower reaches of the River Don to deal with exceptional flows of water. Following the flooding the Ouse Water Board carried out extensive improvements downstream on the River Don, while Bentley Colliery Co. continued to improve the drainage network above its workings.

In the late 1930s the Don Drainage Commissioners, Bullcroft Colliery and Bentley Colliery got together with a view to providing a comprehensive drainage scheme. The idea came to fruition in 1939 and 1940, when Bentley entered into two agreements with the Don Drainage Commissioners to carry out a number of schemes as agents of the Commissioners. Between 1939 and 1945 the Bentley Colliery spent over £70,000 on the schemes. The work was to bring enduring benefit to the colliery and its environs.

It was well known that the Barnsley Seam at Bentley would be liable to spontaneous combustion and the release of immense quantities of gas, which would have to be removed by adequate ventilation. Due consideration of these potential problems was reflected in the layout of the roadways and workings. The underground layout was divided into six, self-contained districts of equal size. Each district had two intake and return roads and its own split from the ventilation, and was thus capable of being sealed off from the rest of the mine in the event of an explosion or fire. Dividing the workings in this manner proved advantageous in later years, contributing to the rapidly increasing output and, when disaster finally came, fulfilling its primary purpose. Bentley became to be universally accepted as the 'Safest pit in the Doncaster district'.

The pillar of coal left to protect the shafts and surface structures of the colliery was 850yds in diameter, containing approximately 120 acres of coal. When the edge of the shaft pillar was reached in 1909 extraction of the Barnsley Seam was commenced by continuous longwall faces, removing the upper 4ft 6in to 5ft 6in of the 9ft thick seam, in order to leave the inferior coal in the upper part of the seam to form a safe roof. In the period 1930-48, in parts of the workings, it was possible to work 18-24in of this top coal and still leave a safe roof.

In the early days production was concentrated in the south, south-east and south-west of the mine, in an attempt to get a substantial output to off-set the large capital expenditure to date. Within months of the commencement of longwall working, difficulties with spontaneous combustion were experienced, due to the immense quantities of methane given off by the coal measures. By 1912, after only two years of working, some forty fires had occurred.

In January 1912 Mr Fryar and Sir Arthur Markham MP, the Chairman of Brodsworth Colliery (Bentley's neighbour, three miles to the west, which suffered from the same problems) were in conclave, with the result that Mr T.H. Bailey, Mining Engineer of

Birmingham, was appointed to investigate the underground fires and report back making recommendations. Bailey reported back on 18 April 1912. In July 1912 Markham and Fryar extended their collaboration and appointed Dr I.S. Haldane (who had been a member of the Royal Commission on Coal Mines, which had preceded the 1911 Coal Mines Act), 'to investigate gob-fires and the method of dealing with them'.

An explosion at Cadeby Colliery, on 9 July 1912, caused the loss of almost 100 lives, including that of Mr W.H. Pickering, the Divisional Inspector of Mines, and several officials in his party, who were caught by a second explosion during rescue operations. The explosion, coming so soon after the 1911 Coal Mines Act, appears to have stampeded the Mines Inspectorate into vindicating its position in the event of a similar occurrence elsewhere and showed the urgency of the task at Bentley. A report to the Home office by the Mines Inspectorate, dated 31 August 1912, noted that the mine was in a highly dangerous condition, and that several serious contraventions of the 1911 Act had been committed.

In an attempt to reduce the risk of underground fires and explosions, a number of measures were taken, including a thorough stone-dusting of the roadways. However, problems were not resolved to the satisfaction of the Mines Inspectorate, with the effect that Summonses were served upon John William Fryar, General Manager, Robert Clive, Colliery Manager, and Albert Longdon, Undermanager. A hearing took place at the Doncaster West Riding Police Court on 16 November 1912. The charges were 'that the pit was improperly ventilated and that men were not withdrawn'. Following evidence from all parties, and further hearings, the case was dismissed on 4 January 1913.

Barber, Walker & Co., having gained engineering prominence through the innovation and success of their sinking at Bentley, had now risen to pre-eminence in the Yorkshire Coalfield, following the defeat of the Home Office at the trial. Had the Inspectorate's interpretation of the 1911 Act been sustained, Bentley, Brodsworth and Bullcroft Collieries might well have been closed for good. In addition, Hatfield, Askern, Harworth and Markham Main Collieries, which were sunk between 1912 and 1925, might never have been sunk.

It was decided to set up the 'Doncaster Coal Owners' Committee (Gob-Fire Research)', with collieries in the Doncaster area subscribing to the Committee. The subscribers would be entitled to copies of reports. Many tests were carried out at Bentley, both underground and in the laboratory. Dr. Haldane's first report presented to the Committee was printed in January 1914. Although the report was the confidential property of the Committee, comments about its importance appeared in the press and, for many weeks after, Fryar was inundated with requests for copies from all over the world. Ultimately, 1,500 copies were distributed. Much original research was carried out by Haldane. In the meantime Fryar intensified the scientific investigation into stone-dusting at Bentley and the Doncaster Coal Owners' Committee proceeded to establish the Mines Rescue Station at Wheatley, Doncaster.

Between 1912 and 1930 Bentley's record was one of solid achievement. In these years the colliery achieved record outputs. Only Brodsworth Main, with three shafts, exceeded these achievements. In 1924 the colliery produced its record output of 1,205,609tons.

During this period the colliery's working costs suffered exceptional charges related to surface drainage. Because the colliery workings were overlain by flat, low-lying land it was necessary to keep a team of men to maintain drainage dykes, embankments and pumping stations. In later years, as the workings extended, surface drainage was to assume even greater proportions and cost.

In 1918 the building of a Baum washer was commenced and new machinery was added to the workshops, which made the colliery almost independent of outside help for major repairs. In 1925 a mixed-pressure turbo-generator was installed to meet the increasing power load. In 1928 a grit-arresting plant was installed by Fraser & Chalmers. This was probably the first in Britain. The plant removed dust and grit from the flue gases.

On 29 November 1931 disaster struck when an explosion occurred in the North-East District of the Barnsley Seam. Forty-five lives were lost and four men were injured. In the inquest which followed, it was found that the disaster was probably the result of a defective flame safety-lamp and a roof fall, in a nearby waste, releasing firedamp. In the immediate aftermath of the disaster many acts of heroic bravery were enacted and on 16 February 1933, at an investiture at Buckingham Palace, His Majesty the King awarded the Edward Medal to seven men for their heroic conduct at the disaster.

The early 1930s were marked by a world-wide economic depression. Bentley's position was made more difficult by the River Don Floods. As a result, only essential improvements were carried out at the colliery. To protect the Don a pillar of coal, some 500yds wide and 3.75miles long, was left beneath the river.

An Elmore, vacuum-flotation plant, the first in this country, was brought over from Aachen, in Germany, and erected at Bentley in 1934. This enabled coal particles, called 'fines', to be removed from washed slurry and used for boiler fuel at the colliery, with a considerable saving to the company.

Following the 1931 explosion, the company set about systematically enlarging the return airways, to reduce ventilating pressure, the liability to spontaneous combustion and gob-fires. After many years of planned effort, many miles of return airway were, literally, large enough to accommodate the proverbial 'carriage and pair'.

On 12 October 1937, with some 26m tons of Barnsley Seam coal still to be worked, deepening of the No.2 upcast shaft commenced, with a view to proving the Parkgate and Thorncliffe Seams (the seams above the Barnsley being of inferior quality). By 11 March 1939 this had been completed. The shaft was now 887yds deep. Brick insets were formed at the Parkgate and Thorncliffe Seam levels, as well as a stone return drift, 20yds above the Parkgate level. The work of sinking continued during the afternoon and night shifts, as well as at the weekends, while coal winding from the Barnsley Seam was carried out on the morning shift. Sinking debris was removed at the Dunsil Seam level and used as packing material in the Barnsley Seam workings. The total cost of sinking was £31,032, or a little over £32.50 per yard and provided an additional reserve of 59m tons of coal. The Parkgate Seam, at 840yds, and the Thorncliffe Seam, at 862yds, proved to be of good quality. The Parkgate had a section of 5ft and the Thorncliffe section was 4ft 8in. The No.1 downcast shaft was not extended, possibly due to the onset of the war, and the No.2 shaft was abandoned and allowed to fill with water to just below the Dunsil Seam, where the level was controlled by a submersible pump.

The declaration of war, in September 1939, caused the company to abandon its plan to develop the Parkgate Seam. It was realized that the delivery of shaft equipment might be seriously delayed, and any development that did not show an early return might be abandoned. For these reasons it was decided to work the Dunsil Seam. This was an inferior seam, but it was less than 20yds below the Barnsley Seam and was workable from the Barnsley pit-bottom. Drifts were driven down to the Dunsil from the Barnsley Seam, and work commenced on the first face in October 1942. Working in the Dunsil Seam, employing mech-anized faces, proceeded rapidly and was very successful.

Bentley and Rossington Collieries were the first in the country to introduce flameproof diesel locomotives for underground haulage. The company ordered a locomotive from the Hunslet Engine Co. and set about preparing the underground roadway to make it suitable for locomotives. The first project was to ride men to and from the district in which they worked, and was operational by August 1939. The effect was to increase the output of the district by about 0.5tons per face worker per shift. Three further locomotives were ordered, together with man-riding carriages, and were operational by mid-1943. By May 1946 eight locomotives were in service at the pit, fulfilling the joint role of man-riding and coal haulage. There was an immediate saving in the replacement of rope haulage systems and, as the success of diesel locomotive haulage became known, the company received many requests for inspection visits from coalfields in Britain and overseas. The company was highly satisfied with the scheme and ordered a further six locomotives in October 1946.

In February 1944, the colliery was paid a surprise visit by King George VI and Queen Elizabeth, who toured the surface and offices.

The successful combination of locomotive haulage and mechanized faces in the Dunsil Seam led to their adoption in the remaining Barnsley Seam Districts. The last hand-filled face in the Barnsley Seam ceased work in December 1945. In total, the cost of this comprehensive modernization scheme was £263,896. For more than a year before nationalization all workers at the colliery travelled to and from their place of work in locomotive-drawn carriages, and all output was taken from the various loading points by locomotive. The only remaining rope haulage was on the incline between the Dunsil and Barnsley Seams, where the gradient was too steep for locomotives.

In 1953 a drift was started to provide locomotive haulage to the east, No.1 area of the Barnsley Seam, where the greatest proportion of the Barnsley reserves were situated. In 1958 the colliery celebrated its fiftieth Jubilee, having provided employment for 3,000 men and produced almost 50m tons of coal. In 1959 the pit-bottom was reorganized and mechanized. A new coal preparation plant was erected and became operational in February 1962. In 1968-69, eight-ton capacity skips were installed in the No.2 shaft, and a new, radial flow fan replaced the existing Capell fan.

With the rapid exhaustion of the Barnsley and Dunsil Seams by the 1970s, it became necessary to seek and develop new reserves. To this end a series of surface and underground boreholes were drilled to prove the extent, thickness and quality of the Parkgate and Thorncliffe Seams. From the results it was decided to deepen the No.1 shaft and to clean out the No.2 shaft.

In 1975 coal sales were rationalized and the preparation plant was replaced by a single washery. The majority of the output went to the CEGB, with a small amount sold for general domestic consumption.

In 1978 the underground conveying was automated and computer-controlled. An entry was begun into the Swallow Wood Seam during 1980, the first production face becoming operational in October 1982. By mid-1986 an £18million scheme, to deepen the No.1 man-riding and materials shaft to the Parkgate level, was completed. The No.2 shaft having been sunk to prove the Parkgate in the late 1930s. A further £4million was earmarked to develop the Parkgate Seam. By mid-1987 the first face in the Parkgate Seam was being prepared for production.

In the financial year ending April 1992 the colliery achieved its first million metric ton output since 1924. Bentley was reaping the benefit from a £30million investment in the

Parkgate Seam and a further £20million investment in heavy-duty face equipment. However, things were not going well for the colliery and the Northern Area Group Director described Bentley as an 'unfortunate victim of the reduced demand for British Coal in a highly competitive marketplace'. At the time around three-quarters of the colliery's 20,000tons average weekly output was being put to stock because of the inability to find a market for it. At the reconvened review meeting, held on 15 November 1993, it was reported that the colliery stocks exceeded 600,000tons and the situation was worsening by the week. Since April 1993 stocks had grown by 396,000tons and the colliery had been purchasing adjacent green-field sites to use as stocking grounds. It was reported that, 'Despite good operational performance, it is clear that there is no prospect for obtaining any additional sales for Bentley'.

On 3 December 1993, with a labour force of 450 men, the colliery ceased production. The colliery had been profitable, but the rundown of the industry due to the reduced market for coal, and the high stocks of coal, sounded the death-knell for the colliery.

Seams Worked:

Seam	**Depth**	**Thickness**
Barnsley Seam	622yds	9ft
Dunsil Seam	642yds	5ft
Swallow Wood Seam		
Parkgate Seam	840yds	5ft
Thorncliffe Seam	862yds	4ft 8in

Brierley Colliery

Brierley colliery was situated ten miles south-east of Wakefield, nine miles south-west of Pontefract, and one and a half miles due north of Grimethorpe Colliery. The colliery was served by both the LMS and the LNER.

In 1912 two shafts were sunk by the Hodroyd Coal Co. to the Shafton Seam, at 217yds. The No.1 was the downcast and coal-winding shaft, the No.2 upcast shaft was used for winding men only. During the colliery's life the Shafton was the only seam of coal to be worked.

In 1919 the colliery was taken over by the Carlton Main Colliery Co. Following the takeover considerable changes were made to the colliery, including electrification of the colliery plant and the provision of electricity and compressed air from Grimethorpe Colliery. Initially, electric power had been supplied by the Central Power Station adjacent to Frickley Colliery. At the same time the screening plant at Brierley Colliery was dispensed with and a surface tramway constructed between Brierley and Ferrymoor Colliery, 1½ miles to the south on the Grimethorpe Colliery site. This enabled Brierley coal to be transported to the screens at Ferrymoor, which had been enlarged to accom-modate the output from the two collieries.

By the early 1940s the Shafton Seam was anticipated to have only a few more years of workable reserves. The seam was 5 to 5ft 6in thick, including a 1ft thick dirt band, and the full section was worked. Towards the north-east of the royalty the seam split rapidly, deteriorating as it approached the boundary of the colliery's take. Water on the faces was a further problem

Brierley Colliery, early 1900s.

at the colliery. By 1945 Brierley Colliery was being phased out and the workforce transferred to Grimethorpe, Ferrymoor and other local collieries.

The colliery officially closed in January 1947, with 233 men on the books, never having been operated by the NCB.

Seams Worked:

Shafton Seam, 220yds deep, 5ft 6in thick.

Brodsworth Main Colliery

Brodsworth Colliery lay in an agricultural area four miles north west of Doncaster, to the west of the Great North Road and the mining village of Woodlands. The colliery was served by three sections of the LNER; the Great Northern, the Great Central and the Hull & Barnsley Railways.

Due to the high capital cost of the undertaking, sinking and equipping of the colliery was undertaken jointly by the Hickleton Main Colliery Co., and the Staveley Coal & Iron Co., of Chesterfield, in Derbyshire. Each company had a fifty per cent share in the new venture, and the Brodsworth Main Co. was registered in 1905 with a capital of £300,000, to work an area of approximately 8,000 acres of coal. The chairman of the new company was the eminent engineer and industrialist Mr Arthur Markham, later Sir Arthur Markham, who was responsible for a number of major mining developments in Yorkshire and South Wales.

The colliery was to become the largest in Yorkshire, and following the sinking of the third shaft, became the highest output three-shaft colliery in Britain. By the early 1920s Brodsworth was claiming the world record for coal drawing, with a daily output of 6,027tons in February 1924, and a weekly output of 30,246tons in December 1923.

Cutting of the first sod took place on 23 October 1905, and sinking of the 20ft 6in diameter No.1 downcast and No.2 upcast shafts was completed to the Barnsley Seam in 1907, at a depth of 595yds. Both shafts, 50yds apart, were equipped for coal winding. During sinking fourteen seams of coal, nine of which were over 2ft in thickness, were passed through. Prior to sinking the permanent headgear, manufactured from oak and pitch pine, had been erected. The intention had been to remove the shaft pillar. For this reason the headgear was manufactured from timber, as this would be more able to withstand the movement arising from the extraction of the coal around the shafts.

Coal production commenced in 1907, when 57tons were drawn up the upcast shaft on 26 November. The output for the first week amounted to 243tons. This coal was produced from the headings being driven in the shaft pillar to form the pit-bottom and main roadways.

The Barnsley Bed was a gassy seam and liable to spontaneous combustion. In preceding years fires and 'heatings' had been common, particularly in those areas of the pit where the Day Bed Seam came to within a few feet of the Barnsley Seam.

The original intention was to use steam for the main winding-engines, while using electricity for all other purposes, both above and below ground. However, early experience in the Barnsley Seam convinced mining engineers that for safety reasons it was necessary to modify this proposal in the vicinity of the working faces and return airways. Therefore, the use of electricity was confined to the main haulages, with compressed air employed on the auxiliary haulages and pumps.

Screening plant was erected 120yds from the shafts in order to reduce the amount of dust drawn down the shaft by the ventilating current.

In common with other collieries in the predominantly rural Doncaster coalfield, the colliery was remote from towns, with only a scattering of small villages nearby. Accommodation for the large numbers of workmen required for the colliery was an issue. In 1907 the company commenced the construction of Woodlands Model Village. By 1917 the village comprised 960 houses, with a population of 6,600. It covered an area of 120 acres, and was surrounded by a further forty acres of shrubbery and plantations. Woodlands Hall, an old mansion in the village, was converted into a workmen's club and institute, and churches and chapels were built. Though the village was a significant outlay, the company regarded it as an investment, enabling them to get the best workman. By the early 1940s the number of houses in the village had increased to 1,647.

An ample supply of red marl clay was available at the site and a brick-making plant, with a capacity of 30,000 bricks per day, was constructed.

In 1916 a Luhrig washer was installed to enable slack (small-coal) to be washed. In 1925 it was replaced with a Baum washer. Introduction of machine-mining in the 1930s resulted in an increase in the proportion of small-coal and dirt, and in 1944 a larger Baum washer, with a capacity of 300tons per hour, was installed.

In 1919 the Brodsworth Main Co. became a member of the Doncaster Collieries Association. The Association was a purchasing and selling organization, and the companies subscribing to the Association retained their separate identities.

Between 1918 and 1920 the No.1 and No.2 shafts were deepened to the Thorncliffe Seam, at 839yds. A third shaft, a 20ft 6in diameter, downcast shaft, was sunk between 1922 and 1923, to the Dunsil Seam at a depth of 608yds. The No.1 shaft then became the main winding shaft, and the No.3 shaft became the subsidiary winding shaft for the Barnsley Seam. The No.2 shaft was left for development of the Parkgate Seam, and by early 1926 coal was being wound from the development headings. However, prevailing adverse trading conditions delayed the development of the Parkgate Seam and it was not until 1933 that the first longwall face was developed in the seam. The Parkgate Seam initially proved to be an excellent seam but, as the workings extended, washouts and problems with the roof affected productivity. In the same year, 1933, machine mining was introduced into the Barnsley Seam.

In 1935 the timber headgear was replaced. In January a steel structure replaced the headgear at the No.1 downcast shaft, and the No.2 upcast shaft headgear was replaced in September. In March 1937 pithead baths, capable of accommodating 3,750 men, were opened.

Formation of the Doncaster Amalgamated Collieries, in 1937, resulted in the establishment of an important colliery company grouping. The amalgamation saw Brodsworth Main, Hickleton Main, Markham Main, Yorkshire Main, Bullcroft and Firbeck Collieries brought together as one group. The amalgamation also brought Brodsworth under Mr W. Humble, an eminent mining engineer of his day, and saw the appointment of a number of key positions at the colliery, which ushered in a period of new prosperity. Brodsworth remained a member of Doncaster Amalgamated Collieries until nationalization.

In the early 1940s the Barnsley, Dunsil and Parkgate Seams were being worked. Barnsley and Dunsil coal was drawn up No.1 and No.3 shafts, and Parkgate coal up the No.2 shaft. The Barnsley Seam suffered from difficult roadway conditions, probably due to the weak roof, and spontaneous combustion was a problem, especially in the north and east of the royalty. The Barnsley Seam had a section varying from 8ft 2in to 9ft 11in, of which 5ft to 6ft 10in was worked. The Parkgate Seam varied from 4ft to 4ft 8.5in thick, and the full seam section was worked. This was a difficult seam to work, because of the irregular nature of the roof, which was treacherous in places.

In 1941 the first face was opened in the Dunsil Seam. The Dunsil Seam section varied from 5ft, at the shafts and in the north, to 2ft 9in. in the southern boundary of the take. The full seam section was worked. Development of the Dunsil enabled the high extraction rate of the Barnsley Seam to be reduced, so extending the life of the seam. In 1948 working in the Thorncliffe horizon commenced.

In 1953 mechanized loading at the coalface was introduced and, by 1960, sixty-three per cent of output was power loaded. The total output of the colliery up to 31 July 1955 amounted to some 47,454,009tons, and the pit was still setting records. In April 1957 the *Yorkshire Post* reported:

Brodsworth Colliery makes mining history.
Record production of 34,231tons with thirty minutes to spare.
Men coming to the pithead landing stage at Brodsworth Colliery, Doncaster, on Saturday, were all asking: 'Well have we done it?' When the answer was 'Yes', they all knew that once again Brodsworth was the biggest pit in Britain. That word of affirmation from the pithead banksman told them that for the first time in British mining history a pit – their pit – had produced more than 34,000tons of coal in one week. The actual output figure of 34,231tons was reached with only thirty minutes to spare before the end of the voluntary shift, but it beat by 393tons the pit's previous best - that of Christmas 'Bull Week' last December.
...Brodsworth, which turned out 30,340tons in a week as long ago as 1924, will eventually be able to advertise itself as 'the biggest pit in the world.

The output record of April was short lived and on 1 June 1957 the *Yorkshire Evening Post* reported:

> ***YORKSHIRE'S 'KING' UP NEW RECORD***
> *BRODSWORTH has done it again! The 3,600 men who work at Yorkshire's 'King Pit' produced 34,422tons of coal during Whitsuntide Bull Week and increased their own national weekly output record. This was the 34,231tons raised during the Easter Bull Week earlier this year.*

In 1956 a major reconstruction scheme was begun, costing £5 million. The aims of the scheme, undertaken in three phases, were to increase productivity and output to 1,680,000tons by 1965. The project provided for the concentration of winding at two levels, Barnsley and Dunsil coals from a new pit-bottom at a depth of 633yds, and Parkgate and Thorncliffe coal from a new pit-bottom in the Thorncliffe Seam at a depth of 850yds. New pit-bottoms were constructed at each of these levels and locomotive roads were driven from the pit-bottoms to link with existing haulage roads, which themselves had been enlarged, to suit 100hp diesel locomotives hauling 3ton-capacity mine-cars. The coal was then to be skip-wound up the three shafts.

The first phase, completed during a three-week period in 1956, converted the No.3 shaft to mine-car winding, winding two 3ton-capacity mine-cars at a time, from the new pit-bottom at the Barnsley level. Phase two, completed in 1960, saw the 900tons per hour coal preparation plant constructed, erection of the No.1 tower-mounted 2,700hp Koepe friction winder and equipment, and reconstruction of the shaft and connecting haulage roadways in the Thorncliffe and Parkgate Seams. The final phase of reconstruction, completed in 1961, comprised the erection of the No.2 tower-mounted 2,700hp Koepe friction winder and

Brodsworth Main Colliery, after closure in July 1991.

refurbishing of the shaft for skip-winding, using 22ton-capacity skips, and was completed in 1961. Following reconstruction productivity doubled from 1.46 to 3.03 metric tons per manshift. On 21 January 1960 the *Yorkshire Post* reported:

> ***PUSH-BUTTON MINE AT A COST OF £3.3m***
> ***Mechanisation plan races ahead***
> *BRODSWORTH – the Queen's pit – will soon be the queen of pits, with kingly treatment for its 4,000 miners and push button mining methods that could mean the housewife will be the first person to handle the coal put in her coal cellar.*
> *...He will walk from the pithead baths along enclosed gangways and descend the pit in the biggest cage in the world, holding 140 men.*
> *The trend is continuing on the surface, where in the twin 158ft high winding towers integrated electrical relays give high speed winding of the coal at the touch of a button. The engineman can have the 22-ton-capacity skip hauled to the surface, discharged and returned to the pit-bottom in just three minutes. The multi-rope skip, operated by 2,700hp electric winding gear, is the biggest in Europe.*
> *Mr Hulley (NCB General Manager for the Doncaster Area) stressed that the entire project had been carried out not only without loss of production but at a time when production continued to increase.*

In 1957 the pithead baths were extended to accommodate 4,000 men, the number calculated to be required to give an output of 2m tons per annum, and at the same time a new medical treatment centre was added.

During the first fifty years of its productive life the colliery produced an average annual saleable output of over on 1m tons. In the 1960s Brodsworth Main was designated as one of the ten high-priority collieries in the Yorkshire coalfield, which were earmarked to receive preferential treatment with respect to manpower, finance and materials. As such, the colliery was destined to play a key role in the activities of the Doncaster Area. At the time the colliery was working the Barnsley, Dunsil and Thorncliffe Seams and had reserves estimated to be sufficient to last for more than a century. In the mid-1960s the colliery was rated the most modern pit in Europe, and supplied coal for Buckingham Palace.

When the Doncaster by-pass was constructed in the 1960s a large quantity of the colliery spoil heap was removed to make way for the road. Subsequently, the spoil heap was landscaped and grassed-over.

In June 1970 two teams of tunnellers linked Bullcroft and Brodsworth collieries with a huge tunnel. The quarter of a mile long link enabled coal from Bullcroft to surface at Brodsworth for washing. In October 1970 the two collieries were merged as a single undertaking, and the whole of the output from Bullcroft was transported underground for winding at Brodsworth. The 700 men from Bullcroft were then transferred to Brodsworth. During the 1970s things went well for Brodsworth, and the colliery produced more than 1.25m tons output on a number of occasions.

In the mid-1970s the No.3 shaft was re-equipped to wind from the Barnsley level, using 12ton-capacity skips. Vertical bunkers were installed at both winding levels (the Barnsley/Dunsil horizon and the Parkgate/Thorncliffe horizon) and were fed by coal from the face, transported via conveyor. Locomotive and mine-car transport was used for materials and man-riding. Two rapid-loading facilities catered for the washed smalls and power station fuel products, each facility incorporating stockpiling facilities. In June 1978 the first retreat mining face at the colliery started production in the Dunsil Seam, as a replacement for the Barnsley Seam.

In the late 1970s the following reserves were estimated at the colliery:

Seam	**Classified Tons**	**Unclassified Tons**	**Total Tons**
Shafton	3,688,000	7,626,000	11,314,000
Newhill	2,142,000	18,424,000	20,566,000
Winter	–	12,999,000	12,999,000
Barnsley	955,000	–	955,000
Dunsil	5,542,000	4,909,000	10,451,000
Swallow Wood	12,309,000	1,274,000	13,583,000
Parkgate	19,162,000	3,430,000	22,592,000
Thornecliffe	11,675,000	–	11,675,000
Top Silkstone	–	4,330,000	4,330,000
Total	**55,473,000**	**52,992,000**	**108,465,000**

At planned output levels the colliery was expected to last until beyond the year 2030. However, by the late 1980s the situation was looking critical for the survival of the colliery. On 9 August 1989 the *Daily Telegraph* reported:

> ***Future of two pits in the balance***
> *The fate of two loss-making South Yorkshire collieries, Markham Main and Brodsworth, and the futures of 1,510 men are likely to be decided this week by British Coal.*
> *The two pits near Doncaster have failed consistently to meet 'survival' production targets set by the company earlier this year.*

On 7 September 1990 the colliery that had once boasted that it was the 'biggest pit in Britain' closed. Eighty-five men had died at the colliery during its working life between 1905-1990.

Seams Worked:

Barnsley Seam, 595yds deep, seam section 5ft 6in to 9ft 6in. thick.
Dunsil Seam, 608yds deep, seam section 2ft 9in to 5ft thick.
Parkgate Seam, 811yds deep, seam section 5ft 4in. thick
Thorncliffe Seam, 842yds deep, seam section 4ft 7in, thickening to the west to 7ft 4in section with 2ft of clod in the middle.

Dinnington Main Colliery

Dinnington Main was located twelve miles east of Sheffield and sixteen miles from Doncaster. It was within eight miles of Worksop, on the outer edge of the North Eastern Division, bordering Derbyshire and Nottinghamshire. Before the coming of mining in the early twentieth century, Dinnington was a predominantly farming community. The colliery was served by the South Yorkshire Joint Line Committee, a branch of the LMS and the LNER.

Dinnington Colliery was originally projected by the Sheffield Coal Co., which, realizing that they did not have the financial resources to sink the colliery, entered into partnership with the Sheepbridge Coal & Iron Co. The resulting Dinnington Colliery Co. was registered on 1 November 1900 with a capital of £187,500, to work an extensive royalty of 7,000 acres.

In 1903 two 19ft diameter shafts were sunk to the Barnsley Seam, which was reached in August 1904, at a depth of 667yds. Production from the seam commenced in 1906. The No.1 shaft was the downcast, equipped for coal winding, and the No.2 shaft was the upcast. The Barnsley Seam suffered from faults and washouts, and in some areas the washouts were attended by double the thickness of coal in the area adjacent to them.

In 1912 coke ovens commenced operation at the Dinnington site using Barnsley Seam coal. They continued in continuous production until decommissioned in 1962. A small brick-plant adjacent to the colliery produced bricks using clay mined locally, and continued in operation until just prior to Vesting Day, supplying the colliery and others in the Amalgamated Denaby Group.

In March 1927 the company was amalgamated with other collieries, and became part of Yorkshire Amalgamated Collieries Ltd. Later in April 1936 the company was incorporated and renamed the Amalgamated Denaby Collieries Co. Ltd, a title kept until nationalization.

By 1933 pithead baths were built by the Miners Welfare. March 1933 also saw the introduction of longwall working, with the coal loaded onto a conveyor. In September 1933 coal was undercut by compressed air saw.

From the late 1940s power loading machines were introduced. First was the Shelton cutter-loader, followed by the Meco-Moore and, eventually, the Huwood Slicer. The latter, introduced in 1953-54, used a prop-free front. In 1964 the colliery became 100 per cent power-loading, employing Anderson Boyes Trepanners.

In the 1940s a battery of twenty 'still' coke ovens at the colliery had a capacity of 4,000tons of coke week. They supplied the Sheffield Gas Co. with 4million cubic foot of gas a day.

Dinnington Main Colliery, February 1992.

Where possible the colliery's output was sent to the coke ovens in order to achieve the highest price for the coal mined.

In September 1960 a large-scale re-organization scheme was initiated, to extend the life of the colliery by concentrating coal working in the little-worked north eastern area of the colliery.

In 1967 the Swallow Wood Seam, lying 80yds beneath the Barnsley Bed, was developed to replace the declining output from the Barnsley Seam. The Swallow Wood Seam was reached by driving two inclined tunnels down from the Barnsley Seam. However, output from the Swallow Wood Seam proved disappointing and in 1971 the Haigh Moor Seam was developed. The last face in the Swallow Wood Seam ceased production in October 1974 and the Haigh Moor soon became the sole seam worked at the colliery.

From the mid-1960s until the late 1970s the colliery experienced declining output and subsequent cash losses. Manpower also declined while output per man shift remained static.

At the beginning of the 1970s the future of Dinnington was examined by the National Headquarters Area Working Party, and in late 1973 the decision was taken to reconstruct the colliery in order to maximize its output to supply the market for coke. Reconstruction commenced with the electrification of the No.1 shaft winder and installation of skip-winding. The scheme cost £1.1million and was completed in August 1975, increasing the shaft capacity by 1,500tons per week. The second stage of the reconstruction was the replacement of the old coal preparation plant, built in 1926 to handle output from the Barnsley Seam. The new plant enabled coal to be prepared for the coking market. It was completed in October 1977, at a cost of £6.7million. The bulk of the coal produced was supplied to the British Steel Corporation (Chemicals) plants at Orgreave and Brookhouse. The final stage of the reconstruction was the electrification of the No.2 winder and installation of new boiler plant. This stage of the project was completed in March 1978 at a cost of £600,000.

Following the completion of the new coal preparation plant, the results for the financial year 1978/9 showed a dramatic turn around, with a profit of £3.4million compared to a loss of £700,000 in 1976/7. In the early 1980s the Haigh Moor Seam was opened up by driving two underground drifts. At this time the colliery was supplying coal to the British Steel works at Scunthorpe, where it was converted into coke.

Dinnington suffered severe deterioration during the 1984-85 National Coal Strike. The Haigh Moor Seam was almost worked out and future production was to come from a new area in the Swallow Wood Seam, to the east of the shafts, where the pit's future reserves lay. Following the strike the roadways being driven to access this new area of reserves were found to need considerable repair work.

By the late 1980s Dinnington's output was being used as a 'sweetener' for power station fuel produced by other collieries in the area. However, having been under threat of closure for a number of years, Dinnington finally closed in September 1991.

Seams Worked:

Barnsley Seam, 667yds deep, some 4ft 3in of the seam section was worked.
Swallow Wood Seam
Haigh Moor Seam

Frickley Colliery

Sunk by the Carlton Main Colliery Co. Ltd, who retained ownership of the colliery until nationalization, the colliery was situated approximately seven miles north west of Doncaster, close to Hemsworth, and within easy reach of the towns of Barnsley, Doncaster, Wakefield and Pontefract. The colliery was served by the LMS Railway.

Because of its central position the colliery shared common boundaries with a large number of surrounding collieries. It bordered Upton to the north, Hickleton to the south, Brodsworth and Askern to the east, and Houghton Main, Grimethorpe and South Kirkby to the west.

Sinking commenced at the beginning of March 1903 and the Barnsley Seam was reached, at a depth of 668yds, on 24 May 1905. On 8 September 1905 some 90tons of coal were dispatched by rail from Frickley Colliery to Grimsby. Two 23ft diameter, brick-lined shafts were sunk to the Barnsley Seam and extended towards the Dunsil Seam in order to establish sumps, at a depth of 670yds. The No.1 shaft was the downcast and coal winding shaft. The No.2 shaft was the upcast shaft, which was used principally for winding men, materials and dirt. Both shafts were equipped for coal-winding and, as the output increased, both shafts were employed to raise coal from the Barnsley Seam. The Barnsley Seam was 9ft 6in thick, and 6ft of this section was worked.

The colliery was equipped with workshops: a wagon shop measuring 150ft by 65ft, blacksmiths and fitting shops, and a powerhouse. In 1918 a brick-making plant was under construction. An interesting feature of the colliery was that every underground district was provided with telephones, connected to an exchange at the pit-bottom and another at the surface.

Owing to the great distance from the nearest hospital, it was felt imperative that facilities for surgical treatment were provided in the vicinity of the colliery. The Warde-Adlam Hospital was founded close to the colliery, through the generosity of Mr W.W. Warde-Adlam, of Frickley Hall, and the directors of the Carlton Colliery Co. The hospital opened in November 1911.

The proximity of the collieries owned by the Carlton Main Colliery Co., namely Grimethorpe, Frickley, Brierley, Ferrymoor and South Elmsall, enabled them to be formed into a group for the purpose of electric power supply. Electric power was supplied at 10,000V from a central power station located adjacent to Frickley Colliery. Electricity was first used at Frickley Colliery in 1906. The power station also supplied power to other collieries in the area including; Upton, Houghton Main and Hickleton Main, as well as local villages, and was connected to the National Grid.

Until 1942, when the Dunsil (or Barnsley Deep Seam) was entered, the total output of the colliery was obtained from the Barnsley Seam. The Barnsley Seam was hand-got until 1934, when mechanical conveyors were introduced. By 1937 face-conveyors were installed throughout the pit. The Dunsil was first entered in a small area to the south of the shaft pillar. Following the initial entry, further access was obtained in 1942 via drifts from the Barnsley Seam, and areas to the east and west were entered. Coal winding remained at the Barnsley level.

At nationalization Frickley was the largest output colliery in the newly formed No.4 Area, raising 4,000-4,500tons per day. The colliery was in good order and the only concern was that Barnsley Seam reserves were seriously depleted, not being expected to last beyond 1967. Planning was then carried out, with a view to deepening the shafts to the Parkgate/Silkstone series.

In 1951 mechanized coal-getting was introduced into the Dunsil Seam and, subsequently, into the Barnsley Seam. Frickley colliery led the way in mechanization with the famous 101's face in the Dunsil Seam. This was a 200yd-long face which was equipped with German-type

Frickley Colliery, April 1994.

friction props, an armoured conveyor with a double-jib coal-cutting machine followed by a fender-plough known as the 'Currie Plough', named after its inventor who was a foreman fitter at the colliery. During its life the face produced 752,150tons at an efficiency rate of 158.7cwt per face worker shift. This system of working was well suited to the difficult and heavy roofs met within the area, and soon spread throughout the No.4 Area.

Frickley was in the big league of coal producers. On 20 December 1956 the *Yorkshire Evening Post* reported, '**1m tons at Frickley for third year**'. While on 29 December 1956 the *Yorkshire Post* reported:

> ***Fourth pit passes million-ton mark***
> *Frickley Colliery last night became the fourth pit in Yorkshire to produce 1,000,000tons of coal this year and the second, a week after Barnburgh, to raise it all up one shaft.... The millionth ton came up the shaft at 5.45 p.m.*

This was the third consecutive year that the colliery had produced 1M ton from the Barnsley and Dunsil Seams, and this figure did not include the 0.25m tons raised from the Shafton Seam, at South Elmsall Colliery, which was run as a separate mine by the NCB.

Skip-winding was introduced into the No.1 shaft in 1964, during a reconstruction scheme. The skips were upgraded to 10tons capacity in 1965, and later to 12tons capacity. The Barnsley Seam was subject to spontaneous combustion, major fires occurring in 1961 and 1966. In 1966 production ceased in the Barnsley Seam.

In 1968 Frickley and South Elmsall Collieries were combined into a single unit under the control of a General Manager. In June 1968 a rapid-loading, 'merry-go-round' system, one of the first in the country, was commissioned. Over half of the colliery's output was dispatched to Ferrybridge 'C' and Thorpe Marsh power stations. By the 1980s the two giant power stations, Ferrybridge 'C' and Eggborough were taking ninety-seven per cent of the pit's output.

By the mid-1960s the colliery was working the Shafton, Dunsil and Barnsley Seams. In order to ensure that its weekly output of 30,000tons did not suffer through lack of reserves, particularly as the Barnsley and Shafton Seams were close to exhaustion, it was decided to drive a pair of drifts from the 225yd deep Shafton Seam to the Cudworth Seam, at 330yds. Two, 560yd-long drifts were driven, at a cost of £200,000. In 1970 the Shafton Seam ceased production, and was replaced by output from the Cudworth Seam. In August 1975 a major project for a new coal preparation plant was begun.

A survey of the combined colliery's reserves, published in the mid-1970s, showed the following:

Seam	**Thickness (inches)**	**Classified tons (millions)**	**Unclassified tons (millions)**
Cudworth	46	7.0	1.2
Newhill	42	5.6	–
Meltonfield	40	4.9	–
Beamshaw	37	8.0	1.3
Barnsley	72	2.5	1.2
Dunsil	65	9.5	0.8
Top Haigh Moor	42	5.6	–
Totals:		**43.1**	**4.5**

It was said that the 'Reserves will take the colliery well beyond the year 2000'. Production from the Top Haigh Moor Seam commenced in January 1984.

In November 1985 the *Coal News* reported:

Frickley Colliery, South Elmsall is receiving a £29million injection to open up reserves in the Meltonfield and Newhill Seams, breaking out of existing shafts at the mid-point 400 metres down. Drivage teams are now heading away from the shaft side and full coal production is four years away.

In March 1986 output from the Dunsil Seam ceased. In 1988, with the colliery producing an output of about 1m tons, eighty-five per cent went to the CEGB for electricity generation, fourteen per cent for coking and industrial use, and one per cent for concessionary uses.

In the late 1980s the colliery was working the Top Haigh Moor and Cudworth Seams, and had a planned 1990 production target of 1.05m tons, from a labour force of 830.

Social and recreational enterprises were well supported at the colliery. They included a top-flight brass band, athletics club, football teams, athletics and boxing clubs. The colliery brass band could be traced back to 1897, when it was known as the South Elmsall Village Band. A popular band, it was in constant demand during the war, for concerts and entertaining the troops.

At the Review Meeting on 12 November 1993 the Northern Area Director said that he had no alternative but to announce the closure of the colliery. He went on to say:

Currently 300,000 tonnes of coal is piled up at the colliery and two-thirds of this has accumulated since April this year. In spite of achieving a modest profit in four of the last five years, Frickley has more recently struggled to produce its coal at costs that meet the demands of a fiercely competitive market.

In the first six months of the financial year 1993/4 the colliery had lost £1.8million, and the production costs were amongst the highest of any British colliery. The Cudworth Seam, which with the Top Haigh Moor Seam, had produced most of the colliery's output, was nearing exhaustion at the time.

On 13 November 1993 the *Daily Telegraph* reported: 'British Coal yesterday signaled the expected closure of South Yorkshire's previously reprieved Frickley Colliery near Pontefract with the loss of 740 jobs.' The report went on to say that production would cease on 26 November, because there was no buyer for sixty per cent of its output. Production ceased at the colliery on 26 November 1993, with the loss of 740 jobs.

Seams Worked:

Shafton Seam, 236yds deep, 1923-1970
Cudworth Seam, 309yds deep, 1970-1993
Barnsley Seam, 665yds deep, 1905-1966
Dunsil Seam, 680yds deep, 1942-1986
Top Haigh Moor Seam, 1981-1993

Goldthorpe Colliery

Goldthorpe colliery was located eight miles north-west of Doncaster, close to the main Doncaster to Barnsley road.

An early colliery at the site was mentioned in a deed of 1678, probably sunk to work the Shafton Seam at shallow depth. The colliery was later worked on a large-scale, with major developments taking place under a Mr William Marsden, of Barnsley, who acquired the property in 1752. Marsden sank an additional five pits and erected a steam pumping-engine in 1770. Marsden's Goldthorpe Colliery lasted until 1783. The colliery then closed, not re-opening until 1909-10, when the new Goldthorpe Colliery was developed by Henry Lodge & Co.

The first sod of the new Goldthorpe Colliery was cut on 29 July 1909. By 1910 two shafts were sunk 64yds deep to work the Shafton Seam, the uppermost seam in the coal measures of the area. The Shafton Seam was one of the thinner seams and was first worked in the Dearne Valley at Goldthorpe Colliery. The seam varied in thickness from 4ft 6in to 5ft 6in, and dipped in a north-easterly direction. The two shafts at Goldthorpe were the 14ft diameter No.1 shaft and the 12ft diameter No.2 shaft.

Henry Lodge & Co. also owned Ryhill Main Colliery and, when Ryhill Main closed in 1923, sold Goldthorpe Colliery to the Old Silkstone Colliery Co.

Because of the shallow depth, the workings were developed on the pillar and stall system, in order to provide better support for the surface. In the early days the output ranged from 1,000 to 1,800tons per week. Later, as the workings advanced further to the dip and the depth of the workings increased, the advancing longwall system was used, the first longwall face introduced in 1950. This resulted in a large increase in output which necessitated improvements in materials-handling and man-riding facilities. By 1960 the weekly output was just over 2,000tons.

In view of the small-diameter shafts, which restricted output, it was decided to drive a drift to intercept the workings to the north-east of the colliery. In November 1954 the Bella Drift was begun. Preliminary site preparations were commenced by the NCB in February 1955. The first 312yds of tunnel were completed by open-cut work. The tunnelling was then given out on contract to the Cementation Co. Ltd. The 2,250yd-long intake drift was driven from the surface at a gradient of 1 in 9, to intersect the Shafton Seam workings. The drift was completed in January 1958. During driving of the drift, heavy feeders of water were released from the seam outcrop and overlying, faulted-sandstone strata. Quantities in excess of 2,000gallons per minute had to be pumped from the face of the drift. However, the problem of water was not at an end. In May 1961 water from the overlying sandstone strata broke into the main return roadway, causing a lot of damage and halting production for two weeks. The drift became the outlet for all the coal produced at the colliery. At the same time as the drift was being sunk man-riding facilities were installed underground, and the surface facilities were upgraded, including new pithead baths, lamp room and workshops. By 1958, with three longwall faces operating, the colliery was producing an average output of 4,500tons per week.

Face mechanization was first introduced in 1960 with the introduction of shearer-loaders loading on to armoured face-conveyors.

The *Yorkshire Evening Post* of 16 June 1961 reported:

> ***FAMILY PITS' SMASH THE OUTPUT RECORDS***
> *Doncaster's two smallest pits – Highgate and Goldthorpe, the 'family pits' where fathers, sons, brothers, uncles and nephews work together – are smashing coal production records. Almost the whole of the output from the twins is fed to power stations, such as those at Doncaster and Mexborough... Two problems which make the performance of the 'identical twins' all the more remarkable is that both have water problems and both have to to use their machines on a shorter face than most pits – only 140yards long.*

In 1966 Goldthorpe and nearby Highgate Colliery were combined when an underground link was made between the two units. Coal from Highgate was transported to the surface at Goldthorpe for processing. In the following year a Colliery General Manager was appointed over the combined mine.

Water in the Highgate section of the mine was drained to a common point half a mile to the north west of the filled shafts, from where it is pumped to the surface by automatic, submersible pumps, which were commissioned in March 1969.

The output from the combined mine was prepared for the CEGB market. In September 1968 a 3,000ton rapid-loading bunker unit was commissioned, enabling merry-go-round trains to be loaded with 950tons of coal and turned around within the hour. New coal preparation plant was also commissioned.

Between November 1967 and October 1969 a drift was driven, within the boundary of the pit-yard at Hickleton Colliery, to meet the Goldthorpe/Highgate workings. A pair of new fans were installed at Hickleton and commissioned in January 1970, and the drift served as a common ventilation return.

Goldthorpe Drift Mine, July 1988.

Washing of tip material at the colliery was tried out successfully, producing some 50,000tons of small-coal per annum in the late 1970s, and about 60,000tons in the mid-1980s. The South Access Scheme, to access reserves in the south-east area of the Shafton Seam, together with the construction of a 1,000tons capacity staple-bunker, was completed in July 1982.

On 1 January 1986 Goldthorpe and Hickleton collieries were merged, and the unit became known as Goldthorpe/Hickleton Colliery.

In the late 1980s a major development took place, to access reserves in the Shafton Seam, in the area south of the Northern Don Fault, which became accessible following the closure of Cadeby Colliery in September 1986. Work started in July 1987 with the driving of twin drifts and by December 1988 the first face in the new area was in production. This new development was to replace production in the south-eastern area. By January 1990 the new area was producing the total output of the colliery. The development had cost £5million and the new reserves were expected to extend the life of the colliery by five to ten years.

In the 1980s Goldthorpe Colliery was a profitable pit, described in 1988 as the South Yorkshire Area's 'little gold mine'. The colliery was producing Britain's cheapest coal from the 54in thick Shafton Seam. At a cost of eighty pence per gigajoule, or £17.81 per metric ton, Goldthorpe was undercutting the price of foreign fuel by more than £2 per metric ton. This cost compared well with the cost of production at the UK's best opencast mines at seventy pence per gigajoule, which was beaten by Goldthorpe one record-breaking week when it achieved a cost of sixty-eight pence, and was the first British pit to go below the seventy pence level. The next lowest output cost at the time, for a deep mine in the UK, was achieved by nearby Frickley Colliery at £1.06 per gigajoule. In the financial year 1988/9 the colliery produced an output of 1m metric tons, for the second time in its history.

In the late 1980s the colliery was consistently undercutting imported fuel prices, and produced coal exclusively for the power station market. In the financial year 1989/90 output topped 9 metric tons per man-shift, a remarkable performance considering that the Shafton Seam was a 'dirty' seam with poor roof conditions. The planned production for 1990 was 850,000 metric tons from a labour force of 513 men.

Production ceased on 4 February 1994 and British Coal revealed that the colliery was one of seven pits to be kept open on a 'care and maintenance' basis, that could be sold off following privatization of the industry. The colliery had, in effect, closed in February 1994, since the bid to find a buyer for the colliery in the private sector failed. Goldthorpe had previously been on British Coal's October 1992 list of nineteen pits to be retained.

At the time of closing the NUM claimed that reserves at nearby Hickleton, Barnburgh and Cadeby could be accessible from Goldthorpe. A British Coal spokesman pointed out that the colliery only had a limited amount of reserves, 'At the rate we were mining, it (the colliery) had about a year. It was a profitable little pit but had few reserves left...' On 19 October 1994 the *Daily Telegraph* reported: 'Two mothballed pits are to be closed after an unsuccessful attempt to find buyers, British Coal announced yesterday. Goldthorpe, near Doncaster, and Kiveton Park, near Rotherham, both South Yorks, will be sealed and demolished.'

The colliery closed for good and by April 1995 all surface buildings had been demolished. In its last year of production the colliery, with 410 men, had produced 1.4m tons of coal.

Seams Worked:

Shafton Seam, 65yds deep, good house and steam-raising coal.

Silverwood Colliery

Silverwood Colliery was located three miles east-north-east of Rotherham, at Thrybergh. Sinking of the two, 21ft-diameter, shafts commenced in April 1900, by Dalton Main Collieries Ltd, who retained ownership of the colliery until nationalization. The No.2 East Pit (upcast shaft) reached the Barnsley Seam, at 740yds, in December 1903 and coal winding commenced at the No.1 West Pit (downcast shaft) in October 1905. The No.1 shaft was used for coal winding and the No.2 for winding men, dirt and materials. During sinking a heavy feeder of water was met at a depth of 150yds. A separate 12ft diameter shaft was sunk, to a depth of 150yds, to divert the feeder and supply the colliery with water.

The colliery had an impressive range of workshops for joiners, fitters and blacksmiths, housed in a building some 180ft long by 40ft wide, as well as a separate wagon-repair shop and sawmill. In 1909, soon after sinking, the colliery was reputed to employ the largest number of men in the Yorkshire coalfield, with 2,593 men underground and 635 above ground. It was probably the largest colliery working a single coal-seam in Britain at the time.

Two batteries of coking ovens were installed at the colliery, the No.1 battery of thirty-six Simon Carves by-product ovens, and the No.2 battery of forty-four Simon Carves regenerative type ovens. In 1912 the two plants were in operation and a tar distillation plant was being erected. By-products recovered from the ovens included ammonia sulphate, ammonium chloride, tar and benzol. An adjoining brickworks had a capacity of 20,000 bricks per day.

The colliery twice received royal visitors, King George V in 1912, and Queen Elizabeth II and Prince Philip in 1975.

When the shafts were sunk at Silverwood, headings were driven from the nearby Round-wood Colliery, to enable Roundwood coal to be wound at Silverwood. Round-wood coal continued to be wound at Silverwood until production ceased at Roundwood in June 1931. Ventilation at Silverwood was assisted by a fan at Roundwood Colliery, some three miles away.

Silverwood had an extensive 'take', about six miles in length and three miles in width, lying to the south-east of the South Don Fault. The general dip of the take was to the north-east and varied from a gradient of 1 in 15 to 1 in 30. The area was relatively undisturbed and what faults were present were generally small.

The Barnsley Seam provided all of the output until 1952, when the Meltonfield Seam, at a depth of 594yds, was developed. During this period annual output often exceeded 1m tons, achieving a maximum of 1,322,501tons, from both shafts, in 1929. The Barnsley Seam was in great demand from the railways because its steam-raising qualities made it suitable for firing locomotives.

In 1953, following nationalization, a major reconstruction scheme commenced. The scheme cost some £3million and was not completed until 1962. The reconstruction involved totally electrifying the colliery, installation of skip-winding, locomotive haulage using mine-cars, new pit-bottoms and transport reorganization, and a new coal-preparation plant. A new, 3,700hp, double-drum winder was installed at the East Pit, to wind coal from the Meltonfield Seam and the Barnsley Seam, using 12ton-capacity skips. At the West Pit a new 2,100hp winder was installed, to wind men, materials and dirt.

Underground, a new pit-bottom for skip-winding was constructed to develop the Meltonfield Seam, and diesel locomotives and 5ton mine-cars were introduced to transport the coal underground. A new pit-bottom was also constructed in the Barnsley Seam and electrically powered trolley locomotives and 5ton, drop-bottom mine-cars were introduced.

A number of problems faced the colliery in the 1960s. These included difficult mining conditions, the poor quality of the Meltonfield Seam, faulting, ventilation problems, the risk of spontaneous combustion in the remaining remote workings of the Barnsley Seam, and the reduction of available reserves. These issues, combined with the need to increase the output and profitability of the colliery, brought about the development of the Swallow Wood Seam. The development scheme was approved in 1965 and production commenced in 1967. For a period of time the Barnsley, Meltonfield and Swallow Wood Seams were worked simultaneously, in order to maintain continuity of output while production in the Swallow Wood was being built up.

On 8 February 1967 Mr Robens, the Chairman of the NCB, visited the colliery. He was doing a tour of the fifty long-life pits in Britain. Silverwood was one of ten such pits nominated in the Yorkshire Coalfield. At the time a major reconstruction, costing about £3million, had just been completed and work was in hand to open up the Haigh Moor Seam. Two, 320yd-long drifts were being driven from the Barnsley to the Haigh Moor Seam, and roadways at the Barnsley level were being enlarged to handle output from the Haigh Moor. The first face in the Haigh Moor was planned to be operational at the end of 1967.

The Meltonfield Seam was finally abandoned in late 1968, having deteriorated, become very 'dirty' and experiencing difficult mining conditions due to the poor roof. The Barnsley Seam was finally worked out and abandoned in 1972. Future concentration was to be in the Swallow Wood Seam. In 1975 a £2.5million scheme was undertaken to develop a large block of Swallow Wood coal in the Braithwell area of the colliery take. The Swallow Wood Seam produced a prime-quality coking coal and three-quarters of its output went to the giant British Steel Corporation, Anchor complex, at Scunthorpe, most of the balance to power stations, with a small amount for domestic consumption.

Further developments took place in the Swallow Wood, beginning in 1976. Combined with the provision of rapid-loading facilities at the surface in 1978, the colliery achieved an output in excess of 1m ton. The initial face in the Swallow Wood Seam was the South Yorkshire Area's inaugural 'spearhead' unit. These were high-output faces designed to produce 1,500tons of saleable coal per day. Plans were made to obtain the colliery's output of 1m ton per annum from four spearhead faces in the Swallow Wood. In 1982 the Haigh Moor Seam was developed, all the faces planned on the retreat system of mining.

In the 1980s the bulk of the collieries output (96 per cent) was supplied to the CEGB via the rapid loading facilities. During this period Silverwood was a boom pit. In early 1989 the colliery raised the fastest 1m metric tons of coal in the South Yorkshire Coalfield's history, and was producing some of Britain's cheapest coal. The colliery was reaping the benefits of the £15million investment in heavy-duty, high-technology faces in the Swallow Wood Seam.

In April 1992 a £10million scheme was announced to access reserves in the Parkgate Seam. The colliery had just had a good year, achieving an output of 1.4m metric tons, its highest ever, and productivity levels were twenty per cent above the national average. The plan was to drive two, 1 in 4 gradient drifts, to access over 11m metric tons of high-quality Parkgate coal by 1995. However, the development scheme did not get off the ground and Silverwood Colliery closed on 23 December 1994, with 378 men on the books. The Swallow Wood Seam was the last to be worked at the colliery.

Seams Worked:

Meltonfield Seam, 572yds deep, 1953-1968
Barnsley Seam, 741yds deep, 1905-April 1972, 5ft-7ft working section.
Swallow Wood Seam, 821yds deep, 1967-1994, 5ft 6in-6ft 6in working section.
Haigh Moor Seam, 836yds deep.

Silverwood Colliery, May 1985.

10
Mining 1911-1920

Despite the First World War, the period 1911-1920 saw the greatest expansion of the Doncaster area of the coalfield. The drive eastwards was unabated and the collieries being sunk were larger, employed more men, were more widely scattered and worked ever larger royalties. These collieries were sunk to work the Barnsley Bed Seam at depths of up to 900yds, with royalties of up to 11,000 acres and, in one case, an option increased the take to 14,000 acres.

The Maltby Trough, having been pierced by Silverwood Colliery in 1905, was pierced again in 1911, this time by Maltby Main Colliery.

In the early years of the twentieth century output from the South Yorkshire region of the coalfield outstripped that from West Yorkshire. 1913 was the peak year for production from British collieries, with an output of 287.4m tons. The output from the South Yorkshire Coalfield was some 27m tons in this year, but it did not achieve its peak until 1929, when it produced 33.5m tons, some thirteen per cent of national output. This delay was because the large collieries being sunk during the First World War did not come into full production for some years. The sinkings in this region of the coalfield were both deeper and technically more difficult, due primarily to the water-bearing rock and quicksand encountered during sinking. Some of the epic sinkings in British coal mining were encountered in the Doncaster region of the coalfield. In addition, the area was predominantly small with a scattering of rural communities, which required the construction of colliery villages and supporting infrastructure, all of which took time.

Collieries sunk in this phase include:

Askern Main
Bullcroft Main
Hatfield Main
Maltby Main
Monckton Main Nos 3 and 4
Rossington Main
South Elmsall
Thurcroft Main
Yorkshire Main

Askern Main Colliery

Askern Main Colliery was situated seven miles north of Doncaster, on the Lancashire & Yorkshire Railway. When the colliery was sunk its layout and equipment was described as the last word in good mining practice. Bounded only by the workings of Bullcroft Main, to the south, Askern's boundaries were largely arbitrary in the early years. To the east the take was limited by ventilation requirements and the economics of the length of the haulages for men, materials and coal.

The Askern Coal & Iron Co. was registered in 1910, to acquire and develop the extensive mining options at Askern, which had been granted to the Don Coal & Iron Co. Ltd. Ownership for the colliery was retained by the company until nationalization.

The initial take comprised 7,000 acres, with a further 7,000 acres available. Because the colliery's workings were sited beneath a predominantly agricultural area it was possible to achieve total extraction, as no areas of coal were needed to support surface features.

After boring to a depth of 375ft, in order to prove the top coal measures, the sinking of two, 21ft 6in diameter shafts commenced in April 1911. During sinking a flow of water was encountered at a depth of 18yds, which increased to 3,000gal. per minute at a depth of 86yds. The initial 60yds were heavily watered and tubbing was inserted through the waterlogged ground. Sinking in both shafts proceeded to the Barnsley Bed Warren House Seam, the top section of the Barnsley Seam. The Warren House was reached in September 1912, at a depth of 566yds in the No.1 downcast shaft, and 564yds in the No.2 upcast shaft. Sinking had proceeded exceptionally well, the maximum rate achieved being 29yds 2ft in a week, a figure believed to be a world record at the time.

Towards the end of 1912 opening-out of the Warren House was begun, while sinking continued until the Flockton Seam was reached in 1913, at a depth of 723yds in the No.1 shaft and 720yds in the No.2 shaft. Both shafts were intended for coal winding, and duplicate sets of horizontal, tandem-compound winding-engines were installed, built by Yates & Thom Ltd, Blackburn. The engines were amongst the largest steam-powered winding-engines ever installed at a colliery.

As development of the Warren House Seam progressed it became apparent that the shafts were sited in a heavily faulted area. This geological uncertainty was to cause considerable difficulty in the development of the colliery. The intention was to develop Askern into a large output colliery like Brodsworth, but the severe faulting, the worst in the Doncaster Area, prevented this being achieved. The Warren House Seam was worked continuously throughout the life of the colliery.

Development and working of the Flockton Seam commenced in 1913 and continued until its abandonment in 1928. The Flockton Seam workings were in the immediate vicinity of the shafts and were connected by drifts to the Warren House Seam. The seam had a section of 3ft of bright coal, overlaid by an inch of dirt and 4in of coal. The floor of the seam consisted of 5in of batt, or inferior coal, and 3.5in of bright coal, giving a total seam thickness of 4ft 1.5in. Approximately 3ft of the section, including the major part of the 3ft of bright coal and part of the batt, was worked. This enabled a safe roof to be maintained. During this early period the Warren House was the principal seam worked. The No.1 shaft was known locally as the Warren House Pit and the No.2 shaft as the Flockton Pit.

In the early days the method of working was by longwall faces, divided into 35yd-long tub stalls, from which the coal was hand-got. From here the coal was loaded into 10cwt capacity tubs and hauled by compressed air haulages to the main haulage roads, from where rope haulage systems were used to transport the tubs to the pit-bottom. This technique of working was followed by advancing longwall units, employing coal-cutters to undercut the coal to a depth of 6ft. The hand-got coal was loaded onto shaker-pan conveyors and loaded into 10cwt capacity tubs. In 1933 rubber conveyor belts were introduced on the coalface, and in 1942 Cardox shells were used to break down the undercut coal prior to hand-filling onto the face conveyor belts.

In the early 1940s the weekly output target at the colliery was 11,728tons, but the colliery was having problems due to the frequent faulting. At this time there was a Coalite plant adjacent to the colliery.

Askern Main, during sinking and equipping of the colliery.

In 1949 the Flockton Seam was re-entered and worked until its abandonment in 1960. Mechanization was introduced into the Flockton Seam in 1958, in its latter years of development, when an armoured face-conveyor and Anderton shearer-loader were installed.

In 1963 a reconstruction scheme costing £589,000 took place at the colliery. The scheme provided for underground drivages and extended trunk-conveying to a 350ton-capacity shaft-bunker, installation of 10ton-capacity skip-winding at the No.2 shaft, and a 750ton-capacity run of coal bunkers at the surface. A short, pit-top tub-circuit and an Aerex radial-flow fan and fanhouse were erected at the top of No.1 shaft. Skip-winding was completed by January 1964, the fan was commissioned in February 1966 and the remaining installations by October 1965. On the week ending 17 December 1966, the colliery broke its record for output of saleable coal, producing 22,123tons in the week.

By this time the Warren House Seam was worked in two main areas of extraction, the Pollington and Smeaton districts. Developments in 1967 resulted in the closure of the Pollington district and concentration in the Smeaton district. The coal was cut by shearer-loader and fed by armoured face-conveyors to gate-belts, which fed a vertical staple-bunker at the bottom of No.2 shaft. The bunker fed 10ton-capacity skips which were drawn up the No.2 shaft.
Underground man-riding and materials haulage to the working faces was by diesel locomotive. At a later date some use was made of man-riding conveyors, to transport men from the outlying parts of the mine to the vicinity of the locomotive haulage.

In the early 1970s the electricity industry provided the major market for the colliery, taking fifty-nine per cent of production. The domestic market took thirty-four per cent and the remaining seven per cent was workmens' concessionary coal.

Further investment in the colliery during the 1970s consisted of a new surface stockyard, improved dirt disposal facilities and improved underground ventilation including a new underground booster-fan. The pit-bottom shaft-bunkerage was also increased. Investment continued and new offices and stores were completed in October 1980, followed by a new coal preparation plant and electrification of the winding-engines in September 1981.

In the late 1980s the colliery was working the Warren House and Barnsley Seams, with a labour force of 470 men. It had a planned production target of 520,000 metric tons in 1990, but the prospects for the colliery were not looking good, as losses had mounted over the years. On 23 November 1991 the *Daily Telegraph* reported:

> ***Loss-making pit to shut at Christmas***
> *Askern Colliery near Doncaster is to close at Christmas when the current coalface is exhausted. The pit, which employs 460, has lost £32 million in the last 11 years.*
> *Mr Alan Houghton, British Coal Selby Group Director, said: 'Askern is an unfortunate casualty of the tough market place we are facing. There is little chance of it being able to compete in a market place which was likely to get even tougher.'*

On 20 December 1991 the colliery, which had the reputation of producing the best house-coal in Britain, ceased production and closed.

Seams Worked:
Meltonfield Seam, 414yds deep, 4ft 3in thick including partings.
High Hazel Seam, 464yds deep, 2ft 6in thick.
Barnsley Warren House Seam, 566yds deep, 13ft thick.
Flockton Seam, 723yds deep, 4ft 2in thick including partings.

Bullcroft Main Colliery

In April 1908 the Bullcroft Main Colliery Co. Ltd was registered, with a take of 11,000 acres. Sunk at Carcroft, five miles north-west of Doncaster, the colliery was served by the LNER, via their Hull & Barnsley and West Riding branches.

Sinking of the two, 16ft 6in diameter shafts, commenced in November 1909 and was completed, to the Dunsil Seam at 684yds, in 1911. Initial sinking was difficult due to the heavily-watered, lower magnesium limestone strata encountered just beneath the surface. Feeders of water amounting to 5,500 gallons per minute were found in the shaft, between a depth of 3-100yds. To complete the sinking through the water-bearing strata, the strata around the shafts were frozen to a depth of 100ft. As sinking proceeded the shafts were lined, with cast iron tubbing to a depth of 128yds, and with brick to the Dunsil Seam. The shafts passed through the Barnsley Seam, at 659yds, in December 1911, and the pit-bottom was formed at the Dunsil Seam level. To protect the surface buildings at the colliery a shaft pillar, measuring 750yds square, was left unworked. Prior to the First World War the colliery was producing 25,000tons per week from three shifts.

Both shafts were equipped for winding coal. The duplicate winding-engines were contained in engine-houses positioned at either end of a central powerhouse. With the exception of the steam winders, the whole of the plant at the colliery, both above and below ground, was electrically driven. It is interesting to note that a large number of integrating meters were installed at the colliery, so that power consumption in each section of the colliery was known and comparisons could be made. Screening plant was erected by Messrs Plowright Brothers of Chesterfield, and comprised three self-contained units. The coal washing plant, consisting of two duplicate units, each capable of treating sixty tons of coal per hour, was erected by the Luhrig company. The whole of the spoil from the screening plant and washers was carried away by an aerial ropeway. Workshops were provided at the surface and included a smithy, a machine shop and a saw-mill.

The Doncaster Amalgamated Collieries Co. was formed on 9 February 1937, and Bullcroft Colliery became a member of the new company.

Throughout the whole of the take the Barnsley Seam was particularly liable to spontaneous combustion and the incidence of 'heatings' was a constant problem during the life of the colliery. In 1941 the owners decided to create a Research Department to investigate safety issues, particularly the problem of spontaneous combustion. As a result of the investigations it was decided to adopt pneumatic-stowing of the side-gate packs to prevent ventilation leakage into the goaf. This system was put into operation in 1944 and was very successful.

Originally, only the Barnsley Seam had been worked, but in 1943 working of the Dunsil Seam commenced. Power-loading was installed on a face in the Dunsil Seam in 1945. On 1 September 1951 pithead baths were opened at the colliery.

The reserves at Bullcroft were almost worked out, and underground conditions, including water problems in the Dunsil Seam, led to an earlier than scheduled closure. It was estimated that closure of the shafts and surface operations at Bullcroft would save the NCB some £250,000 per annum. There had been calls for strike action to stop the merger with Brodsworth Colliery, which failed miserably when put to the vote. Soon 'Bully' and 'Broddy' men were working side by side.

In June 1970 two teams of tunnellers linked the two collieries with a huge tunnel. The quarter of a mile link enabled coal from Bullcroft to surface at Brodsworth for washing.

On 28 September 1970 the *Doncaster Evening Post* reported:

BULLCROFT COLLIERY CLOSES DOWN

Miners, faces like chimney backs, emerged from the shaft at Bullcroft Colliery for the last time yesterday. On Monday, the 700 men from 'Bully' will make their way to Brodsworth Colliery – just one mile away.

In October 1970 the two collieries were fully merged and the whole of the output from Bullcroft was transported underground for winding at Brodsworth. By 1974 plans were proposed to convert the ninety acre Bullcroft Colliery site into agricultural and forestry land, at a cost of £158,000.

Seams Worked:

Barnsley Seam, 659yds deep, the seam section worked varied from 6ft 8in to 8ft 7in.
Dunsil Seam, 685yds deep, 1943-.

Hatfield Main Colliery

Situated one mile from Old Stainforth and seven miles north-east of Doncaster, the colliery was one of the most easterly in the Yorkshire coalfield. It was well-placed for the ports of Hull and Immingham, some forty miles away. Served by the LNER, Doncaster to Hull branch, and Hatfield Staithes on the Sheffield & South Yorkshire Navigation Co.'s canal, the colliery had good outlets for its production.

On 16 December 1910 the Hatfield Main Colliery Co. Ltd was registered, with a capital of £300,000, to work a take of 11,000 acres.

Sinking of two 22ft diameter shafts was started in November 1911. The Barnsley Seam was reached in the No.1 downcast shaft in August 1916. Sinking of the No.2 upcast shaft was delayed by heavily water-bearing strata early in the sinking, and was finally completed in 1917. Due to the problems with sinking, the shafts were lined with concrete throughout. At the position in which the shafts were sunk the seams were in a depression in the coal measures, which was accentuated by a 98yd fault which occurred between the shafts.

The ventilation unit, a Waddle ventilating fan, was driven by a uniflow-type steam engine, an unusual arrangement at a colliery.

In January 1927 2,300 men produced a daily output of 2,400tons. In 1927 the colliery was acquired by the Carlton Main Colliery Co. Only the Barnsley Seam was being worked at the time, which provided steam-raising and house-coal, but difficulties were experienced with spontaneous combustion. The new owners decided to develop the High Hazels Seam at a depth of 766yds, which also produced steam-raising and house-coal. An area of two square miles of the Barnsley Seam was being worked, along with four and a half square miles of the High Hazel Seam, yielding 9,700 and 6,700tons per acre respectively. The Barnsley Seam was wound in the No.1 downcast shaft and the High Hazels in the No.2 upcast shaft.

To enable the High Hazels to be worked, an inset was made in the No.2 shaft at a depth of 733yds but, due to the faults, the No.2 pit-bottom could only handle output from the southern area of the High Hazels. The remaining output was directed to the No.1 pit-bottom. Over the years there were many difficulties with this arrangement and it was decided to create a new pit-bottom in the No.2 shaft, 79ft above the old one, at a depth of 694yds. A drift was driven from the High Hazels to intersect the new pit-bottom. This work took two years. The

Hatfield Main Colliery, April 1993.

new pit-bottom was operational in July 1949. The new arrangement was obviously successful, as shown by the output figures for 1949, which were the second highest so far.

In 1930 the Sheffield & South Yorkshire Navigation, the Aire & Calder Navigation, and Hatfield Main Colliery, reached an agreement for a wharf to be built on the Stainforth & Keadby Canal, together with a number of other improvements. In 1931 the new coal staithes opened and a short rail link from the colliery enabled the colliery to despatch coal by canal. Coal was moved in trains of 'Tom Puddings' (compartment boats) to Goole, for shipping, or to Doncaster power station. Hatfield became the only colliery in South Yorkshire to employ Tom Pudding boats. The canal allowed output from the colliery to be water-borne to Goole, Grimsby, and Hull, on the east coast. By the 1940s twenty-five per cent of output was despatched by this means.

In May 1934 pithead baths, capable of accommodating 2,800 men, were opened. By the early 1940s the coal was 100 per cent machine-mined at the colliery. A section of 6ft of the Barnsley Seam was worked. The High Hazels had a section of 4ft 9in, of which 4ft 6in was worked. Large quantities of water were encountered in some areas of the seam.

In 1945-1946 a reconstruction scheme took place at the colliery. Further development was undertaken by the NCB in 1947 and completed in 1950. This added to the reconstruction begun by the former colliery owners. The total cost of the reconstruction was £383,000; it involved the construction of a new pit-bottom in the No.2 shaft in the High Hazels Seam, 78yds above the existing bottom, to bring it level with that in No.1 shaft and facilitate the switching of traffic between them. New roadways had been driven and locomotive haulage of mine-cars was introduced. The roadways leading from No.1 pit-bottom were also enlarged and re-graded. Following the reconstruction output increased from 686,000tons to 800,000tons per annum.

In May 1957 an underground connection was made with Thorne Colliery, and the two collieries amalgamated in 1967, becoming known as Hatfield/Thorne Colliery. In this year the colliery celebrated its fiftieth anniversary of coal production. During this time over 34m tons of coal had been won from the Barnsley and High Hazels Seams.

Between 1968 and 1971 further reconstruction took place at the colliery, costing in excess of £690,000. In April 1968 a new coal-blending plant was completed. In August 1969 skip-winding was installed in the No.1 shaft. In 1969 an underground booster fan was installed in the north-east return. July 1971 saw the commissioning of a rapid-loading scheme, for bulk loading and rapid turnaround of railway wagons. In July 1972 Hatfield broke local productivity records, achieving more than 3tons per man-shift for the first time in its history.

The steam winding-engine at the No.2 shaft was replaced with an electric winder in July 1974, and the No.1 shaft steam winding-engine was replaced in August 1975.

Methane drainage was in operation in both the Barnsley and High Hazels Seams, the exhausted gas released into the return airways.

Barnsley Seam coal was used in power stations, large tonnages going to the NCB's Immingham terminal, for delivery to power stations sited on the south coast of England. High Hazels coal was prized as a high-quality domestic fuel.

In mid-1978 the NCB Chairman, Derek Ezra, visited the colliery and announced that financial approval for a £10.5million coal preparation plant had been approved. The colliery was reported to have 73m tons of proven reserves and things were looking up. The new coal preparation plant was completed in 1980-1981.

In the mid-1980s new roadways were being driven in a £6million scheme to give access to

11m tons of High Hazel coal, and lead on to a further 20m tons of Barnsley Seam reserves. Production in the new area started in the late 1980s. The development was in the north-west area of the High Hazels, which produced a good multi-market coal. Work was also underway improving the roadways at the Thorne side of the take.

In late 1992 British Coal decided to 'moth-ball' the colliery, together with the adjoining Thorne Colliery. At the Reconvened Review Meeting on 18 November 1993 the Group Director recommended that the colliery closed. Hatfield had the highest production costs in the country. It had lost £8.3million in the six months to the end of September 1993, and had produced 188,000tons of coal, less than half of its planned output. The colliery was operating at a cost of £3.36 a gigajoule, against the target production cost of £1.50. The Area Director went on to say that the colliery had a history of poor delivery-performance and that this was the fifth time that the colliery was the subject of a Reconvened Review Meeting since May 1990. The colliery had made an annual operating profit on three occasions in the last twenty-three years, and in the last four years had lost £18.4million. Production was to cease on 3 December 1993, and the 260 men at the colliery voted not to oppose the proposed closure. The colliery closed, to be re-opened in early 1994 after a management buy-out by the British Association of Colliery Management. Hatfield Colliery is still operational, but was experiencing financial difficulites in late 2001.

Seams Worked:

High Hazels Seam, 766yds deep, 4ft 2in to 4ft 6in thick, producing a good quality house-coal.
Barnsley Seam, 846yds deep, 6ft to 6ft 6in thick.

Maltby Main Colliery

Maltby Main Colliery is situated in a rural area, one and a half miles from Maltby Village, eight miles east of Rotherham and seven miles south of Doncaster, close to the boundary between Yorkshire and Nottinghamshire. It is connected to the South Yorkshire Joint Railway line, and served by the LNER from Tickhill Junction. The colliery is in the concealed part of the South Yorkshire Coalfield, overlain by 170ft of Permian limestone.

In 1907 the Maltby Main Colliery Co., a subsidiary of the Sheepbridge Coal & Iron Co., decided to sink two 20ft diameter shafts to work 9,000 acres of the Barnsley Seam. Sinking of the No.2 shaft commenced on 30 March 1908. The Barnsley Seam was reached on 17 June 1910 at a depth of 820yds. On 21 January 1911 the No.1 shaft reached the Barnsley Seam. Water was released by springs within the first 200yds of sinking, amounting to 21,000gallons per hour. Two pump lodges were built in the shaft to deal with the large quantity of water.

The shafts were found to have been sunk on the crest of an anticline (like an upturned saucer), with the seams dipping gently at a gradient of 1 in 20 from the shafts in all directions. In a later reconstruction scheme the shafts were deepened to the horizon of the rim of the anticline to make working easier.

The No.1 shaft was used to draw the whole of the colliery output, while the No.2 shaft was used primarily for winding men and materials.

The working conditions in the Barnsley Seam were among the best in the county, but the seam was prone to spontaneous combustion and was very gassy. Later this resulted in the colliery

Maltby Main Colliery, following reconstruction in early 1960s

being one of the first pits in the Division to employ methane drainage. The seam also suffered from a weak roof. Large quantities of industrial gas and coking coals, and the world-famous London Brights, were mined at the colliery. Until the Swallow Wood Seam was developed between 1968 and 1971, the Barnsley Seam was the only seam worked.

In 1926 the colliery passed into the control of the Denaby & Cadeby Colliery Co., and in 1936 became embodied in the Amalgamated Denaby Collieries Ltd, who retained ownership until nationalization.

On 8 October 1927 the first pithead baths were opened at the colliery, and further improved baths were built in 1939. Initially the coal was hand-filled into 13cwt tubs, from longwall faces, which were transported to the shafts by endless-rope haulage. By 1935 1M ton per annum was being mined in this way. In 1944 compressed air driven conveyors were introduced to transport coal from the face to loading points.

In the 1940s the company owned 1,300 houses in the surrounding villages, and the Maltby Miners' Transport Co. ran a fleet of buses connecting the colliery with the villages of Tickhill, Braithwell and Maltby, where the majority of the workmen lived.

In the period 1947-61 a major reconstruction scheme was undertaken to increase the colliery's capacity to 1.2m tons per annum, initially exclusively from the Barnsley Seam. At the time it was estimated that there were some 36m tons of reserves in the Barnsley Seam. The scheme would make the colliery one of the largest in Yorkshire. The reconstruction entailed the deepening of the shafts, by some 47yds, to the new winding depth from the Barnsley Seam at 892yds, and construction of new pit-bottoms. New main roadways were driven, and skip-winding was installed at the No.2 shaft. One of the first tower-mounted, multi-rope friction-winders in the country was erected over the No.2 shaft. About fifty per cent of the output was hauled underground by diesel locomotives, from trunk-conveyor loading points, and these were replaced by 95hp, electric-battery locomotives hauling 3ton-capacity mine-cars. At the surface a new coal preparation plant was constructed, a new fan was installed and electric winding-engines replaced the two existing steam winders. The No.1 shaft was electrified for winding men, materials and dirt. The No.2 shaft was fitted with a 4,000hp, four-rope friction-winder and 12ton-capacity skips. The reconstruction scheme cost some £3.25million.

Until 1954 coal had been hand-filled onto conveyors, as the large amount of methane had restricted the use of electricity on the coalface. This was one of the main reasons for the delay in introducing cutting machines. In February 1954 the first power loader, a compressed air driven Meco-Moore machine, was introduced. This was followed by the introduction of the Hugh Wood Slicer, which was a very successful type of power loader. The use of this machine and careful attention to grade proportions enabled the colliery to achieve a higher percentage of large-coal output than average in its area (thirty per cent against an average of twenty-four per cent).

Methane drainage was installed at the colliery in 1956, and was gradually extended to all production panels. Initially it discharged to the atmosphere. However, at a later date the gas was pumped into a surface ring-main and piped to Manvers Coking Plant, where it was used for firing coke ovens.

In the mid-1960s the colliery's output was split between the following markets: coke – forty-one per cent; domestic – twenty-five per cent; gas – twenty-two per cent and railways – nine per cent. In late 1968 a new development scheme, to improve output-capacity, was approved. The scheme cost over £1.25million and was to take until 1971 to complete. The scheme opened up the Swallow Wood Seam, to replace the limited reserves in the extensively worked Barnsley Seam, and extended and modernized the coal preparation plant to handle the additional output.

Between 1975 and 1976 work on a rapid loading system, using merry-go-round trains, to speed up the shipping of coal to the British Steel Corporation coke ovens at Scunthorpe, was implemented.

By the mid-1980s ninety-two per cent of the output was sold to power stations, loaded to merry-go-round wagons via a rapid loading system.

A £1.5million boring programme was carried was out in the central and south-eastern area of the colliery's take. This identified reserves of 100m metric tons. In 1981 a project to access these reserves was approved. The scheme, costing in excess of £170million, was to increase output to over 2m metric tons per annum, making the colliery the first 2m metric tons per year mine in South Yorkshire. The scheme involved the sinking of a third shaft, 1,000m deep and 8m in diameter, to wind all of the colliery's output and open up reserves in the Parkgate and Thorncliffe Seams. The reserves in these seams would provide a further 50m tons of prime coal. Sinking of the No.3 shaft commenced in October 1982 and was completed in November 1989. When the scheme was launched in 1981 it was the NCB's largest ever investment at an existing, fully-operational colliery. In 1992 the first face was working in the Parkgate Seam.

In late 1992 British Coal decided to 'moth-ball' the colliery, but in early 1993 decided that it would be extensively developed to enable it to produce at a significantly lower cost in the future. The colliery was to be developed as a high-volume, low-cost unit, capable of producing 2m metric tons of coal per year. Up to £40million was to be spent over the coming two years developing the extensive reserves for retreat mining.

In December 1994, when the surviving coal industry was returned to the private sector, Maltby Colliery was purchased by RJB Mining, who continue to manage and work the colliery.

Seams Worked:

Barnsley Seam, 820yds deep, 5ft to 7ft thick, 1911-1972.
Swallow Wood Seam, 894yds deep, 7ft 10in thick, 1970-1987+.
Haigh Moor Seam, 901yds deep, 4ft 2in thick.
Parkgate Seam, 1,050yds deep, 4ft 9in thick. 1988-

Monckton Collieries Nos 3 and 4

The headings of the old Hodroyd Colliery had extended some 900yds, reaching the boundaries of neighbouring collieries by 1892. Exploiting the Shafton Seam, beneath the Newstead Estate, from the existing pit would have been impractical. The decision was taken to sink a new shaft at a site between Havercroft and Upper Hiendley. Sinking commenced in July 1892, reaching the 5ft thick Shafton Seam, at a depth of 87yds, on 3 November 1892. As the usual practice was for shaft numbers to be assigned to the deeper shafts, this shallow shaft was usually referred to as the Galway or Lady Galway's Shaft.

By the turn of the twentieth century the coal beneath the Newstead Estate had been extensively worked and, as this formed the boundary with Hemsworth Colliery, the Shafton Seam would soon be exhausted and new reserves would have to be developed. The Lady Galway's Shaft, at old Hodroyd Colliery, was chosen as a site to sink a shaft to work the Barnsley Seam beneath the adjoining Newstead and Holgate Estates. The sum of £30,000 was set aside from profits to sink the 16ft diameter No.3 shaft at the site. The intention was that the No.3 shaft and workings would be worked by electricity from the outset, but problems with the newly formed Yorkshire Electric Co. were to dissuade the company for the time-being. On 30 September 1913 sinking started, continuing until interrupted by the general strike in February 1914. On 3 March 1915 the 4ft 8in thick Barnsley Seam was reached at a depth of 568yds. Work then proceeded to drive headings to join up with the nearby Felkirk workings of Monckton No.1 pit. The new No.3 shaft would ensure that Barnsley Seam coal in the Felkirk area would be more accessible and profitable, by reducing the travelling and transportation time. The ventilation was also improved by the installation of powerful, new fans.

At the time a decision was taken to deepen the No.3 shaft to the Haigh Moor Seam, but the decision was vetoed by Lord Galway's agents, possibly because this was the second year of the First World War and times were far from normal. The decision to sink to the Haigh Moor Seam was finally taken in September 1921, when it was decided to sink a new shaft on the site. The early 1920s had witnessed the closure of three local collieries; Hodroyd, Ryhill Main and South Hiendley, due to exhaustion of the Shafton Seam. Together with the success of the No.3 shaft, these factors may have acted as a catalyst for the go-ahead of No.4 shaft!

Sinking of the 20ft-diameter No.4 shaft started on 5 April 1923 and by 21 May 1924 the shaft had reached the 3ft thick Haigh Moor Seam, at a depth of 643yds. Headings were driven into the coal, and a pit-bottom, large enough to accommodate triple-deck cages, was built. A roadway was also driven into the 3ft 4in. thick Barnsley Top Softs Seam, which lay at a depth

New Monckton Nos 3 & 4 Colliery, late 1950s.

of 549yds, in order to link to the No.3 pit workings and assist with ventilation. By November 1924 the roadway was complete and both shafts were in production. The electrification, which had been proposed for the No.3 pit at the outset of sinking, was to become reality by 1925.

In the mid-1920s a large-scale housing development was underway nearby, to alleviate the housing problem which the company was experiencing.

The winding-engines installed at the No.3 pit were second-hand and the makers had warned that the strain under which they were working was too great. In June 1928 one of the piston rods broke, resulting in the colliery standing idle for two weeks. A new engine was ordered from Markham & Co., of Chesterfield, and was delivered on 30 August 1930. Three-deck cages were fitted, in place of the existing ones, to increase output and recoup the additional expense.

In the late 1950s the colliery was mining the Lidgett Seam and the recently mechanized Kent Seam, and looked forward to a long and prosperous future. With reserves of 150m tons the colliery had a projected life of 150 years! In reality the life of the colliery was to be a further seven years.

The NCB had run the two production units, New Monckton Nos 1, 2 and 6 and Nos 3 and 4, in tandem, until closure on 3 December 1966. For the first time the reasons for closure of a colliery were given as economic rather than on the grounds of exhaustion. After initial vigorous union opposition to closure, the recommendations were accepted. At closure New Monckton Nos 1, 2 and 6 employed 1,273 men, and Nos 3 and 4 employed 1,000 men. Following closure most of the workforce were relocated to neighbouring collieries.

Seams Worked:

Shafton Seam, 87yds deep, 5ft thick.
Kent Seam, 406yds deep, 2ft 9in thick.
Barnsley Top Seam, 549yds deep, 3ft 6in thick.
Barnsley Seam, 568yds deep, 5ft thick.
Haigh Moor Seam, 643yds deep, 3ft thick.

Rossington Main Colliery

Located three and three quarter miles to the south-east of Doncaster, off the Great North Road, the colliery was sunk by the Rossington Main Colliery Co., a subsidiary of the Sheepbridge Coal & Iron Co., and John Brown & Co. The company was registered on 15 July 1911 with a capital of £750,000. Rossington Main was served by the South Yorkshire Joint Line Committee, the majority of the traffic travelling by the old Great Northern Railway section.

The two shafts were sunk in an area in which the seams had a general inclination of 1 in 30 to the south-east. Sinking of the 22ft diameter, No.1 downcast shaft was started on 10 July 1912. Between 24 July 1912 and 21 February 1913, sinking had stopped due to the influx of water into the shaft, which varied from 1,800 to 2,100gallons per hour. The shaft reached the Barnsley Seam, at a depth of 873yds, on 3 May 1915. Sinking then proceeded to the Dunsil Seam at a depth of 888yds. Shaft sinking was completed in the No.1 downcast shaft on 16 May 1915

Sinking of the 20ft diameter, No.2 upcast shaft commenced on 26 June 1912, reaching the Barnsley Seam on 7 November 1915, and the Dunsil on 18 November. Sinking finished on 7 December. Having reached the Dunsil Seam, drifts were driven up to the Barnsley Seam and development of the seam proceeded. Coal-winding from the Barnsley Seam commenced from the No.1 shaft on 17 July 1916. 107,800tons were raised in that year. Initially both shafts were fitted and used for coal-winding.

The Barnsley Seam varied from 6ft 9in to 7ft 9in in thickness. Of this, a 5ft 7in section was being worked in the early 1940s. The seam produced a gassy coal and was prone to spontaneous combustion, which necessitated a high standard of ventilation and large roadways. Until 1956 the whole of the colliery output came from the Barnsley Seam.

In 1927 the colliery became part of Yorkshire Amalgamated Collieries Ltd which had a capital of £3.2million and comprised Rossington Main, Dinnington Main, Maltby Main, Darton Main, and the Strafford Colliery Co. Ltd. In January 1933 mechanization arrived when conveyors were first installed in the Barnsley Seam. By the 1940s the colliery company owned some 1,700 houses situated in New Rossington Model Village. In 1953 a reconstruction scheme was initiated, replacing endless-rope haulage with 100hp diesel locomotives, for the transport of coal to the pit-bottom. The scheme necessitated the driving of new drifts, the enlargement of existing roadways and the grading of roadways.

During the mid-1950s a dense-medium washer, the first of its kind in Britain, was designed and erected at the colliery by Simon Carves Ltd. Towards the end of 1956 limited development of the Dunsil Seam took place, to replace output from the depleted Barnsley Seam. A methane drainage scheme was installed at the colliery, being completed in 1962. Exhausted from boreholes in the vicinity of the working faces, methane was piped to a surface exhauster via the No.2 shaft and pumped ten miles to Manvers Main for firing coke ovens.

In 1963 a start was made on the installation of skip-winding in the No.2 shaft. In the same year the construction of new, 158ft high, headgear was started.

In 1964 Nos 1 and 2 shafts were deepened to the Swallow Wood Seam, the No.2 shaft to 927yds and the No.1 to 928yds, to enable skip-winding to be put into operation in the No.2 shaft, and the use of larger cages into the No.1 shaft. In September 1964 No.2 shaft went into operation with 12ton skips, driven by a ground-mounted, electrically-powered friction-winder at the surface. All coal was wound at the No.2 shaft, and No.1 shaft was used for winding men and materials.

Rossington Main Colliery, July 1992.

During 1962-1963 power loading was introduced at the coalface. In June 1964 the last hand-filled face closed. All faces were power loaded, using compressed air powered machinery, except for the Dunsil Seam where electricity was permissible. In September 1965 powered supports were introduced at the colliery and were soon in general use.

In 1965, a 160ft high headgear, with a ground-mounted friction-winder, and a new engine house, were erected on the No.1 shaft, and put into operation in September of the same year.

Coal production in the Dunsil Seam ceased on 18 October 1968 and output then came totally from the more profitable Barnsley Seam. In 1969, following the commissioning of a battery of electrically powered compressors, the pit-top was totally electrically powered.

In November 1969 the first stage of an underground conveyor and bunkering system was commissioned. The scheme was to bring coal directly from the coalface to a bunker at the shaft side for loading into skips and winding to the surface.

In early 1971 a rapid loading facility was completed, in order to streamline the despatch of bulk tonnages by merry-go-round trains, principally to power stations. This was one of the first such systems at a colliery in Britain. A new Coalite plant at the colliery opened on 8 September 1972.

By 1977 methane, which had been supplied to the coking plant at Manvers, was used to fire boilers at the Rossington and Yorkshire Main Collieries.

In the mid-1980s there were still vast reserves in the Barnsley Seam and work was undertaken to develop the east side of the colliery. In autumn of 1989 work started on a new, £5million coal preparation plant, as the old, antiquated plant could no longer handle the pit's output.

In 1993 the colliery was closed by British Coal and offered for sale to the private sector. In 1994 the colliery was leased by RJB Mining, along with two Nottinghamshire collieries. RJB invested in the colliery to re-equip it, and two years later bought the colliery, extending the operating license from ten to twenty-five years. On 22 July 1996 the *Yorkshire Post* reported:

> ***RJB buys out coal mine it held on lease***
> *Coal group RJB Mining has bought Rossington Colliery near Doncaster, after running the mine under a lease and license arrangement for the last two years. RJB has extended the pit's operating license from ten to twenty-five years in a move – good news for the 320 men working there.*
> *Rossington was close by British Coal in 1993 but reopened by private mining group RJB a year later and has since been producing coal for domestic and industrial markets...*
> *Bill Rowell, managing director of RJB's deep mine operations, said it was delighted the long-term potential of Rossington's extensive reserves had been recognized...*

Seams Worked:

Barnsley Seam, 873yds deep, 6ft 9in to 7ft 9in thick.
Dunsil Seam, 888yds deep, 5ft 9in thick.
Swallow Wood Seam, 928yds deep, 4ft 6in thick.

South Elmsall Colliery

South Elmsall Colliery was situated approximately seven miles north-west of Doncaster, close to Hemsworth, within the largest parish of the largest county in England.

The colliery was served by the North Eastern and Midlands Regions of British Railways, which dealt with its total output.

The colliery was originally sunk as Frickley Colliery No.3 shaft, to the Shafton Seam at 249yds. Frickley No.2 shaft served as the upcast, and Nos 1 and 3 as downcast shafts. The colliery was sunk by the Carlton Main Colliery Co. Ltd, who retained ownership for the colliery until nationalization. Sinking of a 14ft diameter brick lined shaft commenced on 14 April 1920, reaching the Shafton Seam, at 249yds, on 7 November 1923. Nine days later sinking was complete, and soon after the shaft was used for coal winding.

Overlain by a strong water-bearing sandstone, of varying thickness, the general dip of the seam was 1 in 20. The Shafton Seam had a section of 4ft 8in and the full seam thickness was worked.

For the first two years of operation the coal was worked by the pillar-and-stall system, using compressed air picks to win the coal. In 1925 the seam was closed, due to quota restrictions and a lack of a market for its coal. It did not reopen until 1942.

Upon reopening, the colliery's name was changed from Frickley No.3 pit to South Elmsall Colliery, and was placed under separate management. Frickley No.2 shaft was used for winding men and materials for both Frickley and the new South Elmsall Colliery. For two years after reopening the faces were worked by longwall retreat. This proved unsuccessful, due to the influx of water from the overlying, water-bearing sandstone and the low yield achieved. Various trials were made with American-type mining equipment, but with little success. Advance longwall faces were introduced in 1945, followed by continuous mining in 1946. In 1953 the entire surface at South Elmsall was reorganized and modernized.

South Elmsall Colliery, April 1994.

Compressed air for use at South Elmsall and Frickley Collieries was provided by Frickley Colliery Power Station, and electricity by the Yorkshire Electricity Board or, alternatively, Grimethorpe Central Power Station.

A Simon Carves, Baum-type coal washer and a froth-flotation plant were in use at the colliery. Coal was produced for electricity generation plants, mixed with Barnsley Seam coal for the house-coal and gas markets, and lump coal was supplied for the house-coal and shipping markets.

By the 1960s the entire production was mechanized. It was delivered by trunk-conveyors into a staple shaft bunker near the pit-bottom, where it is loaded into mine-cars for diesel locomotive haulage to the pit-bottom.

In 1968 Frickley and South Elmsall Collieries were combined into a single unit under the control of a General Manager.

In 1970 the Shafton Seam was exhausted and ouput from the seam was replaced by production from the Cudworth Seam, which was worked by retreat mining.

In September 1971, 4ton-capacity skips were installed on the No.3 shaft. These were increased to 5ton-capacity in July 1982.

South Elmsall Colliery was closed in November 1993, at the same time that Frickley Colliery closed.

Seams Worked:

Shafton Seam, 236yds deep, 4ft 8in thick.

Cudworth Seam, 309yds deep.

Thurcroft Colliery

Thurcroft was situated in a predominantly rural area, five miles to the south-east of Rotherham, and eight and a half miles east of Sheffield. The chosen site had ample room for expansion. The colliery was served by the LMS and the LNER companies.

Mineral rights were leased to the Rothervale Colliery Co. in 1902, but sinking did not commence until 1909. Sinking of two, 20ft-diameter shafts, in close proximity to New Orchard Farm, was undertaken by the Rothervale Mining Co. The shafts were sunk to the Barnsley Seam. No.1 shaft was the downcast and No.2 the upcast. The No.2 shaft was sunk to a depth of 744yds, where a fault running east-west, with a downthrow of 150yds, was encountered. This faulted the Barnsley Seam out of the shaft. An inset was constructed, at a depth of 667yds, and roadways were driven to the rise-side of the fault, to locate the seam. The No.1 shaft was sunk to a depth of 750yds, where an inset was also constructed. From this inset a 1 in 1 inclined drift was driven to the Barnsley Seam, which lay on the dip side of the fault at a depth of 820yds, Connecting roadways were constructed to meet the No.2 shaft. Coal was not reached until January 1913.

In about 1913 the first company houses were built for the developing colliery, to be followed by further developments in 1926.

The Barnsley Seam was extensively exploited in the early life of the colliery, and was to produce the bulk of the output in these years. Workings were to the rise of the fault between 1915-31, and to the dip between 1930-68.

In 1921 the Rothervale Mining Co. were taken over by the United Steel Co., who used the colliery to supply their steel works with coal and coke. The company then became the Rothervale Collieries Branch of the United Steel Co., who were to own the company until nationalization. Following take over by the United Steel Co., a brickworks and coke ovens were built on land adjacent to the colliery.

The colliery reached its peak in the 1930s, when some 2,000 men and boys were employed, and daily output was approximately 2,500tons. Coal was supplied for coking, steam-raising and gas-manufacturing purposes.

Difficulties in working the Barnsley Seam had resulted from the major Thurcroft Fault, as well as a series of small faults running north-west to south-east, on the western side of the royalty, towards Treeton Colliery, with throws of up to 45ft. Working the Barnsley Seam in this direction proved to be unprofitable and it was decided to leave this coal for Treeton to work. In addition there were two large washouts north-east of the shafts in the Barnsley Seam. Working of the Barnsley Seam ceased in 1968, due to the lack of a market for high-quality steam-coal following the changeover of British Railways from steam to diesel power.

Drifts were constructed to the Parkgate Seam and in 1942 development of the seam commenced, providing a high grade coking coal. Working of the seam continuing until abandonment in 1972, due to difficult working conditions and extreme temperatures, because of the great depth.

In 1970 working of the Haigh Moor Seam commenced. All the seams at the colliery provided high-quality coking coal, which had a high demand in the steel industry.

During the Second World War the manpower at the colliery declined. Following nationalization the workforce increased to pre-war levels and two further housing estates were built in the early 1950s, to accommodate the workforce.

Thurcroft Colliery in 8/134.

In 1953 a new fan was installed and some underground reconstruction was carried out. Further reconstruction took place in the 1970s. Electrification of the No.1 winder was completed in 1975, and a new coal preparation plant was commissioned. At the time almost all of the output of the plant, in the form of washed smalls, was supplied to the British Steel Corporation, Scunthorpe works.

The colliery was connected underground, in the Barnsley Seam, to Treeton Colliery. This connection was sealed in 1970 following an underground fire.

In late 1977 the NCB Chairman, Derek Ezra, visited the colliery and announced a £3.2million investment. This was to develop the Swallow Wood Seam, in order to take over from the exhausting Haigh Moor Seam, and boost output of the colliery's high-quality coking coal by 180,000tons, to 525,000tons. This would create 6.3m tons of additional reserves, which would extend the life of the pit by twelve years. In 1983 drifts were driven to the Swallow Wood Seam, on the dip side of the fault.

In the mid-1980s ninety-five per cent of output was despatched to the British Steel coke ovens at Scunthorpe. In 1988, ninety-seven per cent of output was transported by rail.

Geological problems at the beginning of the 1990s were to bring about the closure of the colliery. On 29 November 1991 the *Daily Telegraph* reported:

> ***Thurcroft pit set to close***
> *Massive losses, predicted to hit £15 million by the end of the financial year, caused by severe geological problems, have led to the closure of Thurcroft Colliery, claims British Coal…*
> *Mr Siddall said it was his intention that the 660-man Thurcroft Colliery should cease production next Friday.*

Coal production ended at Thurcroft Colliery on 6 December 1991. This was not quite the end for Thurcroft Colliery, as the colliery received a reprieve in early 1992. On 19 March 1992 a local newspaper ran the following article:

Boost in pit battle

A coal mine which closed with the loss of 650 jobs got a reprieve yesterday.

British Coal agreed to keep Thurcroft Colliery, near Rotherham, in working condition while privatization plans are drawn up. It still has twenty years of coal reserves and could make money, said campaigners, the NUM, Rotherham council and MPs.

In late May 1992 ex-miners were given two weeks to lodge their rescue plan for the colliery, the *Daily Telegraph* reported:

Miners float pit rescue bid

Miners were yesterday given two weeks to lodge a rescue plan to reopen Thurcroft Colliery.

British Coal will switch off the pumps and cap the mine if a group of former employees, backed by Rotherham Borough Council, fails to lodge viable proposals by June 5...

Thurcroft was axed with the loss of 660 jobs just before Christmas. British Coal claim it had lost £9.6million over the previous six months and geological problems made it uneconomic.

A company called Thurcroft Colliery (1992) Ltd was formed and, on 5 June 1992, put in a bid for the colliery. The consortium hoped to buy the colliery in August and commence production in September. A feasibility study came to the conclusion that Thurcroft had access to 20m metric tons of marketable coal and could be made profitable.

British Coal had stipulated that the company meet maintenance costs, while the future of the colliery was decided. The company agreed to carry the care and maintenance costs but were only able to maintain the payments for three weeks. British Coal had previously maintained the colliery for three months while the company prepared its business plan.

A statement from British Coal published in August's edition of *Coal News* spelled out the situation: 'In view of the Company's refusal to meet the care and maintenance costs, British Coal has no alternative but to resume the shaft sealing operations which had been originally planned for completion last March.'

Seams Worked:

Barnsley Seam, 820yds deep, 1913-1968, 6ft 11in thick.
Swallow Wood Seam, 885yds deep, 1983-1991, 48in thick.
Haigh Moor Seam, 899yds deep, 1970-1987.
Parkgate Seam, 1,050yds deep, 1942-1972.

Yorkshire Main Colliery

Located four miles to the south-west of Doncaster, the colliery take was situated beneath agricultural land. The colliery was served by the LMS and LNER. The Staveley Coal & Iron Co., of Derbyshire, were interested in the colliery sinkings being undertaken in the Doncaster area of the South Yorkshire Coalfield. In 1907 they acquired mineral rights in the vicinity of Edlington. In 1909 the sinking of a colliery began. Initially known as Edlington Main, the name was changed to Yorkshire Main in September 1909, and in 1913 the Yorkshire Main Colliery Co. was formed.

Sinking of the two 21ft 6in diameter shafts commenced in 1909 and were completed, to a depth of 911yds, in 1911. The Barnsley Seam was found at a depth of 906yds, and varied between 8ft and 10ft 1in in seam section. Of this 5ft 8in to 6ft 10in was worked. The Barnsley Seam proved to be a very gassy seam, giving off large amounts of methane, and a high standard of ventilation was required for safe working. In addition to the two main shafts, a water shaft was sunk to a depth of 130yds, to provide water for the colliery.

The Barnsley Seam was worked continuously from 1911 until closure of the colliery in 1985. As development took place the shafts were found to lie in a heavily faulted area, which resulted in the pit-bottom lying beneath the major reserves in the Barnsley Seam. Initially, the Barnsley Seam was worked with longwall faces, the coal being hand-filled into tubs at the coalface. In 1933 the tub system was replaced by longwall advancing faces, with conveyors along the coalface feeding to a loading point in the main roadway. In October 1930 the pithead baths were opened, with accommodation for 2,200 men.

The colliery was one of a group owned by Doncaster Amalgamated Collieries Ltd, who initiated a project to reconstruct the colliery. The colliery had been sunk in a faulted area and, after a period of working, it was realized that the major reserves were at a higher level than originally anticipated. Prior to nationalization work had been carried out preparing plans and a limited amount of underground work had been completed. The project initiated by the former owners was approved by the Capital Expenditure Committee of the NCB on 12 September 1947. The scheme consisted of the introduction of underground, diesel-locomotive haulage using large-capacity mine-cars, installation of skip-winding at the No.2 shaft and electrification of the winding equipment and power plant. Improving the locomotive haulage necessitated the construction of a new pit-bottom, 60yds above the original level, and the driving of connecting roadways from existing haulage roads to the new pit-bottom level. Construction of the new underground haulage roads was completed by 1952. Erection of skip-winding equipment began early in 1952 and was completed by January 1953. Mine-cars were gradually introduced and the changeover from tub to mine-car winding at the No.1 shaft was complete by July 1953. The No.1 shaft was reconstructed to wind men and materials, but could also wind coal at times of peak output. The scheme also included the building of new workshops for the repair and maintenance of mine-cars. Following the reconstruction, output rose from 810,000 to 914,000tons. In total the reconstruction had cost £1,118,257.

In 1953 the reserves remaining to be worked in the Barnsley and Dunsil Seams amounted to 82m tons. These reserves were estimated sufficient to give the colliery a life of seventy years at a production rate of 4,500tons per day. Almost the whole of the coal was despatched by rail from the colliery.

Power loading at the coalface was introduced in 1954. Initially mechanization accounted for some sixty per cent of output and further mechanization was planned for the near future. Further improvements to the coal preparation plant were completed in 1955, with the installation of a new washer and water-clarification plant.

In 1962 a methane drainage scheme, for the exhaustion of methane from the underground workings, was completed. The gas collected was piped to Manvers Main for firing coke ovens.

A large proportion of the workforce lived in the adjoining village of New Edlington, where the NCB owned a large number of houses and the Coal Industry Housing Association had built further houses. The village of New Edlington was well provided with facilities. There was a recreation park and a club with cricket and football grounds, tennis

Yorkshire Main Colliery, 1980.

courts and a bowling green, as well as an open-air swimming pool and recreation hall on the colliery premises.

In the early 1970s the Swallow Wood Seam was being developed as a replacement for the Barnsley Seam, which was becoming exhausted. The first face in the Swallow Wood became operational in the first quarter of 1970/1. To allow the seam to be exploited at the required rate, a new return roadway was completed in June 1972.

The colliery never recovered from the National Coal strike of 1984-1985. In October 1985 the *Coal News* reported:

> ***No reason to keep pit open – Tuke***
> *...Explained Mr Tuke (Area Director): 'The colliery review committee has agreed that the pit could and must produce 14,000tonnes a week, compared with the current output of just over 7,000tonnes – a figure which falls short of the target by a long way.*
> *In the financial year so far, to the end of August, the pit has lost £83 a tonne. It has lost a total of £7 million in the five months and continues to lose money heavily. Productivity should have been doubled...*
> *The result of all the factors is that recovery at Yorkshire Main has not been satisfactory and the prospects do not appear to be there. In the light of the failure to achieve the agreed recovery programme, I can see no justification for continued operations at the colliery.'*

Yorkshire Main ceased coal production on Friday 11 October 1985.

Seams Worked:

Barnsley Seam, 907yds deep, 8ft 6in thick, 1911-1985.
Dunsil Seam.
Swallow Wood Seam, 1971-1985.

11
Mining 1921-1939

The final phase of the development of the South Yorkshire Coalfield came in the period 1921-1939, with the sinking of Thorne and Markham Main Collieries in the Doncaster area, and the Upton Colliery in a small area to the north-west of the coalfield.

Sinking at Markham Colliery started in 1916, but ceased soon after for the duration of the First World War. Sinking resumed in 1922, reaching the Barnsley Seam in 1924.

Both Thorne and Upton Collieries proved to be beset by problems in working coal, both having been sunk by companies from the North East. Upton Colliery was sunk as a joint venture between Bolckow Vaughan & Co. Ltd and the Cortonwood Co. Ltd, who later sold out to Dorman Long. Thorne was sunk by Pease & Partners, who suffered severe financial problems as a result of the venture.

In addition two further collieries were sunk at Harworth Main and Firbeck Main, to the east of Maltby Main and Dinnington Main, but these collieries were in Nottinghamshire and are not included within our remit.

Collieries sunk in this phase include

- Markham Main
- Thorne
- Upton

Markham Main Colliery

Markham Main Colliery was situated in the village of Armthorpe, two and a half miles to the east of the centre of Doncaster. The colliery was served by the South Yorkshire Joint Lines Committee's Railway, and came to have an excellent landsale facility. The site selected for the colliery was particularly open, with considerable space for future surface expansion.

In June 1913 the lease to the mineral rights at Armthorpe near Doncaster, was secured from Earl Fitzwilliam by Sir Arthur Markham. Sir Arthur died soon after sinking commenced and the sinking was to become known as Markham Main Colliery, in his honour.

Sinking of the two shafts was started on 6 May 1916, but ceased on 24 August 1916 for the remainder of the First World War. Sinking resumed briefly in mid-1919 but was abandoned until 21 May 1922 when sinking resumed. Work continued until the Barnsley Seam was reached, at a depth of 729yds in both shafts, during May 1924.

The two 17ft diameter shafts were sunk to a depth of 746yds, passing through the Barnsley Seam at 731yds. The No.1 shaft was the downcast, and No.2 the upcast. The shafts were lined with concrete through the water bearing strata, and with brick for the remainder. The Mines Department's Technical Adviser's report on the colliery, dated 6 January 1942, made note of the size of the shafts, considering them to be too small for a modern colliery working a seam at this depth.

Coal winding using 'hoppits' began in May 1924. Double-deck cages carrying tubs were installed in the No.2 shaft in August 1924, and in the No.1 shaft in March 1925. With both shafts equipped for coal winding, a weekly output of 16,000tons was reached by 1926. This increased to 20,000tons by January 1930, and 22,733tons by May 1940. The Barnsley Seam varied from 5ft 10in to 7ft 6in in thickness, and 3ft 6in to 5ft 2in of the seam was worked. In places the seam suffered from wet floor conditions and, in the southern districts, poor roof conditions were a problem and spontaneous combustion was an ever-present danger. Special precautions were taken to prevent underground fires. The Barnsley Seam was worked extensively at the colliery, providing the whole of the output up to 1954.

The steam winding-engines were located at either end of a long enginehouse, and between the winding-engines was situated the powerhouse, housing compressors and switchgear for the colliery.

After the main haulage roads were driven through the shaft pillar, the coal was worked on the long wall system of tub stalls, and was hand-cut and hand-filled into tubs.

A colliery village with 993 houses was completed by 1927. The first house had been occupied in January 1920. Rent, rates and lighting were deducted by the colliery from the weekly wages of the workmen.

Tub-stall working continued until January 1934, when the first conveyor was installed and working by longwall advancing faces was introduced. By June 1942 all coal was transported from the faces by conveyor and fed onto gate-belts.

In March 1937 the first coal-cutting machine was installed at the colliery. In the same year the colliery was taken over by Doncaster Amalgamated Collieries Ltd, who retained ownership until nationalization.

In the late 1930s things were not looking good for the colliery. A report, presented to a Board Meeting in May 1938, stated that the life of the pit, which was working the Barnsley Seam in the area north of the 140yds Fault, was limited to less than ten years. The report went on to set out the advantages and disadvantages of working the coal to the south of the fault from the Markham shafts against sinking new shafts in the vicinity of Auckley or Blaxton. It was anticipated that either scheme would take five years to develop. The submission to the Board was to work the area to the south of the fault by drifting through the fault. It was proposed to sink new shafts to the south of the fault, further to the east of the original proposal, on the Blaxton estate, an area owned by the company. It was argued that by investing £160,000-£180,000 a considerable area south of the fault could be developed and the life of the colliery extended by over forty years. Two pairs of drifts were proposed, the East Drifts and the South Drifts, which would give access to 22.6m tons and 18.8m tons of coal respectively. The East Drifts were to have been completed in 1946 and the South in 1951.

The recommendations of the report, in terms of sinking new shafts, were never implemented. The Second World War and nationalization would have had an impact on any plans formulated.

Pithead baths for 2,016 workmen were opened in May 1938 and a canteen for 300 workmen in August 1943, followed by a medical treatment centre in February 1948.

Man-riding using rope-hauled 'paddy mail' was introduced into the colliery in December 1938 in the North-East District, and introduced into other districts soon after, so that man-riding paddy mails were running over some 8,000yds of roadway. Grading of roadways, for the introduction of locomotive haulage, began in November 1946 and the first diesel locomotives were put into use in August 1947.

Markham Main Colliery, July 1992.

In 1954 yielding hydraulic props were used on the coal face in place of rigid steel ones. Power loaders were introduced soon after. Twenty per cent of output was power loaded by 1959, and ninety per cent by 1965. In April 1966 the first powered supports were introduced into the colliery. To handle the increasing output from the faces, trunk-belt conveying was introduced in 1962, replacing diesel locomotive haulage.

The Dunsil Seam was worked from 1954 on the north-east side of the pit. In 1961 a partial reconstruction scheme was authorized. The scheme included deepening the No.1 shaft to the Parkgate Seam, improving underground and surface transport, and driving a large drift from the south-east area of the pit, where production was concentrated at the time. In 1963 further reconstruction was approved. This included introducing skip-winding in the No.1 shaft below the Barnsley level, sinking a bunker staple shaft from the Barnsley Seam to the skip-winding level, and extending the underground conveyor system to the staple shaft bunker.

The Barnsley and Dunsil Seams were located so close together that, on 28 September 1961, the *Doncaster Chronicle* reported. 'Markham can claim 14ft of coal in a district in which the Barnsley and Dunsil Seams run together'.

In 1980 the NCB estimated available reserves at the colliery as follows:

Barnsley Seam	32.8m metric tons
Dunsil Seam	4.4m metric tons
Parkgate Seam	35.7m metric tons
Thorncliffe Seam	3.5m metric tons
Total	**76.4m metric tons**

A further 25m metric tons of coal were available in thinner seams, such as the Cudworth, Newhill and Meltonfield. At planned output rates the colliery was expected to last until at least 2050.

Electrification of winding was completed in 1980, and by July 1980 winding capacity had been uprated. In the late 1980s work started on a £11.75million scheme to construct a new pit-bottom.

By the early 1990s the prospects for Markham Main were not looking good, and industrial relations with British Coal were at a low ebb. On Saturday 2 May 1992 the *Daily Telegraph* reported:

> ***Walk-outs pit shuts with 730 job losses***
> *British Coal announced the closure of Yorkshire's Markham Main Colliery with the loss of 730 jobs yesterday, the first pit closure since the election.*
> *The company said the sixty-eight year-old mine had lost £30 million over five years, including £2 million since May due to a series of walk-outs in a dispute over the use of private contractors.*
> *In a terse four-minute statement read to officials of the National Union of Mineworkers, Mr Bob Siddall, the Area's Director, said a rescue plan had been 'aborted' and he added: 'I shall be advising the corporation that I can see no future at Markham Main and that the colliery must close.'*

In October 1992 British Coal announced thirty-one closures and the loss of 30,000 jobs, with ten collieries earmarked for immediate closure. One of the ten was Markham Main, which at the time was profitable and had substantial reserves of high-grade coal suitable for the power generation market. Markham Main Colliery was closed on 16 October 1992.

On 3 May 1994 the *Daily Telegraph* reported:

> ***Pit to reopen***
> *Production is to resume at Markham Main, the former British Coal pit near Doncaster, South Yorks, under a deal with the Coal Investments firm.*

Coal Investments, headed by Malcom Edwards, the former British Coal commercial director, leased Markham Main from British Coal. Things did not go well for Coal Investments, who reported losses of £18.2million in its start up year to 31 March 1995. The losses reflected the cost of reopening the five collieries which they had taken over and the company's policy of immediately charging these costs to profit and loss. As a result Markham Main Colliery finally closed in 1996.

Seams Worked:
  Barnsley Seam, 731yds deep, 5ft 10in to 7ft 6in thick.
  Dunsil Seam

Thorne Colliery

Thorne is the most easterly colliery in the Yorkshire Coalfield, situated at Moorends on the low-lying Thorne Levels, which are barely 10ft above the Ordnance Datum Level. Pease & Partners, a well-established company, of Darlington, in County Durham, sank the colliery to provide coking coal for their iron and steel-making plants. The High Hazels Seam was a high-quality clean coal, free of dirt bands, and admirably suited to their requirements.

The two shafts were among the most difficult to sink in the history of the Yorkshire Coalfield, taking longer to sink than any other shafts in the history of mining. Both severe water

Thorne Colliery, late 1920s.

problems and the outbreak of the First World War made the sinking formidable and significantly increased the cost of sinking.

Sinking commenced on 12 October 1909. The initial contract went to Messrs Eaton Son & Hind, of Thorne, who carried the sinking down to the top of the Bunter sandstone. A halt was made at this level to erect the head frames and install sinking pumps and winding-engines. When sinking recommenced large quantities of water were encountered and it was decided to employ the Shaft Freezing Co. to apply the German freezing-system. In December 1909, sinking recommenced under the direction of the German parent company, *Tiefbau und Kaltenindustrie Aktiengesellschaft*. The skilled operatives, and many of the workmen, were German and at the outbreak of the First World War work ceased. Following the end of the war it was decided to abandon the freezing process in favour of the cementation process and Messrs Pease & Partners entered into a contract with the Francois Cementation Co. Ltd (later known as the Cementation Co.). Sinking restarted on 18 March 1919 and was complete on 13 March 1926. The two 22ft diameter shafts were sunk to work the High Hazels Seam at a depth of 862yds. The shafts were lined with cast iron tubbing at the top, and with concrete and brickwork lower down.

About a quarter of a mile away was the colliery brick works. This supplied the colliery's needs, as well as bricks to build Thorne New Village, which had more than a thousand houses, built for colliery workmen. An adjacent clay-pit supplied the raw materials.

Exhaust steam from the winding-engines and fan were used in mixed-pressure turbines in the colliery power station. In addition to power generated for use at the colliery, the power station also provided the new colliery village with electric lighting, as well as supplying power to the Yorkshire Electricity Board's network.

The High Hazels Seam was worked manually until late 1933, when conveyors were installed. In 1952 power loading was introduced, with the installation of Meco Moore cutter-loading machines. The results from the new machines were promising.

At the completion of sinking the shafts were found to be relatively dry, but water was to cause problems in the future. In 1936 'weepers' of water appeared in the No.1 shaft and in 1944 water was found to be issuing from the concrete lining, in the shaft, at a depth of 380yds. Later that year cracks appeared in the No.2 shaft at a depth of 400yds. Repairs were carried out but in April 1953 the flow of water from the patched crack in the No.2 shaft became noticeable, increasing from about 90 to 269 gallons per minute. Further remedial work could not stem the flow and, by April 1954, the flow had risen to 1,400 gallons per minute. Further measures reduced the flow to 540 gallons per minute. Despite these steps, the flow had risen to 1,300gallons per minute by August 1955.

It was established that the fundamental problem was the existence of 'freezing tubes', which were driven through the impervious anhydrite beds. These were relics left from the original attempt to sink using the freezing method. The result was that water migrated down these tubes and penetrated the shaft lining in the underlying Permian strata. The lining had not been designed to withstand these unforeseen pressures. The precise number and position of these tubes was not known as the data holding this information had disappeared when the engineers returned to Germany. Between 1944 and 1955 a search was made for cavities and freezing tubes, many of which were located and sealed. Despite this effort they were unsuccessful in stemming the flow of water. On Tuesday 17 July 1956, *The Times* reported:

> ***WATER CAUSES PIT CLOSURE***
> ***TWO YEARS' WORK ON SHAFT ENVISAGED***
> *Production at Thorne Colliery, near Doncaster, which employs 2,400 men and has a weekly output of 14,000tons, is to cease, probably for about two years, while a large part of the shaft is relined with concrete to deal with inrushing water.*
> *Major-General Sir Noel Holmes, chairman of the North-Eastern Division of the National Coal Board, said today: 'All the Thorne men will be gradually re-absorbed in other collieries in the area...We deeply regret this decision, which has been forced on us by technical difficulties...We expect that the work will take two years and cost about £250,000. But the pit has a life of 80 years and reserves of 99m tons of coal.'*

In December 1956 the NCB suspended coal winding operations at the colliery, in order that effective steps could be taken to repair the damaged wall in the No.2 shaft. The existing lining was strengthened with reinforced concrete, reducing the original 22ft diameter to 18ft in diameter, and to 16ft in diameter over a 60yd section midway down the shaft. It was also decided to insert cast iron tubbing, for a distance of 213yds, between the base of the Lower Limestone and the Triassic sandstone. No.1 shaft was lined with a welded, steel lining, backed by concrete for part of its length. No.2 shaft was repaired in a similar way, over an area in the shafts known as the 'bad patch'. This work was unusual and complex, and was reported on in various mining journals of the day.

It was agreed that while the colliery was closed for these shaft repairs, a full reconstruction, both underground and at the surface, would be carried out. The objectives of the reconstruction were to install skip-winding in the No.1 shaft, to construct underground roads suitable for locomotive haulage, to mechanize the coalfaces and to erect new surface plant for the handling, washing and grading of output. By 1958 approximately half of the reconstruction was complete, and it was expected that the colliery would soon be back on-stream. However, fresh cracks appeared in the No.2 shaft and it was decided that further internal lining and reinforcing of both shafts was required. The shaft repair work was expected to take until 1964, but was not finished until 1966. Relining the shafts had cost £1,660,000.

In 1967 proposals were submitted for reopening the colliery but, due to the contraction of the industry, it was put on a 'care and maintenance' status. The colliery had an underground connection to neighbouring Hatfield Colliery, which served as an air-intake and emergency exit route.

In 1967 a bulge appeared in the No.1 shaft lining, which worsened in 1970. Repairs were carried out in 1973 but, later that year, a second bulge appeared.

In the early 1970s spoil from the colliery tip was removed to form the sub-base for the M18 and M62 motorways, and the M18-M62 motorway intersection, lying to the north of the colliery site.

By 1974/1975 it was accepted that new capacity was needed to replace that lost in the older parts of the coalfield. Reopening Thorne had the advantage of existing access and it could be brought into production more quickly than a 'green field' site. In response to the NCB's desire for primary capacity, Headquarters gave tentative approval for reopening the colliery for coal production. The concept of a new mine based at Thorne looked favourable. The High Hazels had been worked for thirty years, and the Barnsley Seam, which had been proved in the shafts, was unworked. Unusually, the Barnsley Seam had never been explored by boring from the surface, or the High Hazels workings. Therefore all of the workings at the colliery were exploratory and subject to a degree of risk. As a spin-off from the work being done to prove the Selby Coalfield, a number of boreholes were drilled to explore the Thorne take. The boreholes agreed closely with predictions and proved extensive reserves in both the Barnsley and High Hazels Seams. Following the results from the surface bore holes, underground borings were taken from the High Hazels Seam.

The results of the boreholes showed the following reserves (after a reduction of sixty per cent from the 'plan reserves' to cover losses due to working, faults and washouts):

Meltonfield Seam	20,058,000tons
High Hazels Seam	54,073,000tons
Barnsley / Warren House Seam	65,465,000tons
Total	**139,596,000tons**

These reserves were anticipated to provide the colliery with a life of fifty to 100 years, at a saleable output of 1.5 to 2m tons per annum.

The requirement for 'new' capacity resulted in the consideration of Thorne as the site for the development of a 2.1m metric tons per annum operation, employing 1,450 men.

In January 1976 a feasibility study was submitted to Headquarters. In March 1977 a Stage 1 was submitted, with proposals for reopening Thorne Colliery. Stage 1 was approved and in May 1977 authority was granted to establish a project team to prepare a submission for Stage 2 and to undertake the rehabilitation of No.1 and No.2 shafts and associated works. In October 1977 a project team was established at the colliery and work proceeded. In January 1978 a contract was given to Thyssen (GB) Ltd for rehabilitation of the shafts. New winding-engines, enginehouses and headgear were to be erected on No.1 and No.2 shafts, a number of surface buildings were to be demolished and the surface was to be generally tidied up.

Various other contracts were let out for soil-surveys and boreholes. In 1979 additional financial authority was granted for site preparation and equipping of No.3 shaft. In January 1980 the Stage 2 application was approved. The entire scheme cost £212million. The new No.2 headgear was erected during October 1980 and the No.1 headgear was erected during February 1981.

The plan was to sink a third shaft for coal winding and upcast ventilation, while the two existing shafts were to be retained as downcast shafts and used for winding men and materials. The No.3 shaft was to have had four, 18ton skips operated from two, tower-mounted friction-winders. The surface was to have been completely rebuilt with a comprehensive coal preparation plant incorporating facilities for the despatch of coal by road and rail. Output from the Barnsley and High Hazels Seams was to have been prepared as a washed domestic/industrial fuel, with the balance despatched as blended, power station fuel. The mine was to be designed with a high level of automation, environmental monitoring and computer control from a surface control room. Initial output was to have been 0.5m metric tons a year, increasing to 2m metric tons within two years.

On 1 February 1986 Thorne Colliery, which was being run on a development only basis, was merged with Hatfield Colliery under a single management team. In early 1988 the *Coal News* reported:

> ***THORNE IS PUT BACK ON ICE***
> *Proposals for the re-opening of Thorne Colliery – a massive £111 million investment programme which would have created 700 new jobs in South Yorkshire – have been put on ice.*
> *There are three main reasons why the decision to 'mothball' Thorne for the second time was taken by Area Director Ted Horton – the NUM's total opposition to any form of flexible working, poor industrial relations in the Area and the continuing national overtime ban…*
> *The latest plans would have seen Thorne start production again within three years to produce 1.5 million tonnes a year. Longer-term proposals envisaged further massive investment to provide a third shaft, output increased to 4.5 million tonnes a year and more new jobs.*

Since 1979, when the scheme to reopen the colliery had begun, £34million had been spent on new headgear and other surface installations, and a new pit-bottom circuit.

Thorne has been put on a long-term 'care and maintenance' basis, figuring in plans to replace capacity at the Selby Complex, in the North Yorkshire Area, at some point in the future.

Seams Worked:
High Hazels Seam, 862yds deep, 1926-1956, 5ft thick.

Upton Colliery

Situated at Upton, one mile north of Hemsworth, the colliery was connected to the LNER (Cudworth to Hull section), three quarters of a mile away.

Upton Colliery was a joint venture between Bolckow Vaughan & Co. Ltd and the Cortonwood Co. Ltd, who later sold out to Dorman Long, who then retained ownership for the colliery until nationalization. The Upton Colliery Co. was registered in November 1923 with a capital of £600,000. Preparation of the site for the sinking of two shafts, to the Barnsley Bed, was begun on 26 March 1924. The two shafts, the East and West Pits, were situated 60yds apart and were sunk to work an area of 8,000 acres.

Sinking of the West Pit commenced on 26 September 1925, reaching the Barnsley Seam, at 713yds, on 22 January 1930 (sinking had been suspended from 10 August 1927 until 2 January

1930, due to a severe depression in the industry). Sinking to the final depth of 738yds was completed on 4 February 1930. Sinking proceeded normally in the shaft until a depth of 91ft was reached, when a water feeder of 1,200gallons per minute was struck. Other feeders were encountered at depths of 225ft and 300ft. To overcome the influx of water, lengths of tubbing were inserted in the shaft. The Shafton Seam was found at a depth of 300yds and what little water percolated into the shaft was collected in an inset in this seam and pumped to the surface. Where the shafts were not tubbed they were lined with concrete blocks, fourteen to the circumference, which were cast at the surface in a specially designed mould.

Sinking of the East Pit commenced on 26 January 1926, reaching the Barnsley Seam, at a depth of 711yds, on 12 June 1927. Sinking was completed, at a depth of 736yds, on 25 June 1927. Water was not a problem in this shaft and the pit was lined throughout with concrete blocks, similar to those used in the West Pit.

A range of modern workshops at the colliery contained a blacksmith, joinery and fitting shops. Brickworks adjacent to the colliery, served by clay-pits half a mile away, provided bricks for the colliery buildings and workmen's houses. In 1928 a housing scheme proposed the construction of 1,100-1,200 houses, shops, a school and a library to the east of the colliery. The colliery was also linked to the Carlton Main Colliery Co.'s power station at Frickley, from where it received its electric power.

The new colliery suffered problems from the outset and throughout its relatively short life was rarely free from geological problems and frequent outbreaks of spontaneous combustion. The shafts, surface equipment and layout were designed for an output of 1m tons per annum. It was anticipated that the output from the Barnsley Seam alone would reach 1.25m tons per year, but this was never achieved.

When the shafts were sunk a fault was discovered on the edge of the shaft pillar. This had the effect of throwing the Barnsley Seam vertically downwards for 100yds on the south-east side of the colliery take. Since the boundary between Upton and its neighbour South Kirkby was only 1,100yds to the south-west, this meant that two sides of the colliery's take quickly became non-productive, and work proceeded to open out the Barnsley Seam on the north eastern side of the take.

Upton Colliery, early 1900s.

Coal working commenced on 4 July 1927. The Barnsley Seam at the colliery was a high-quality coal, suitable for industrial and domestic purposes. The seam was 14ft 9in thick, but the maximum thickness of coal which could be worked was approximately 5.5ft. The coal was heavily faulted and the continual halting of work to negotiate the hundreds of faults which were encountered, frequently resulted in localized 'heatings' and underground fires.

The north-eastern side of the take had to be sealed off in 1935, due to spontaneous combustion. Further faces were sealed for the same reason, and in 1945 an area of the South Barnsley District was closed. This was a particular blow to the colliery as this was an area which had recently been opened and had been very productive. Over 2,000 men were laid off until October 1945 when the pit reopened.

At nationalization Upton was the 'youngest' colliery in Yorkshire, but since then had never been in a sound financial position due to the many difficulties which it had faced. It was decided that the objective at Upton was to open up sufficient face room so that an economic output could be raised. The Barnsley Seam was being developed, adjacent to the area which had been sealed off due to spontaneous combustion. The Beamshaw and Winter Seams were also being opened up and large size roadways were being driven into a vast area of coal to the east and north-east of the pit. This would keep the colliery going for many years once economic working was established.

Many unsuccessful attempts were made to overcome spontaneous combustion. Larger roadways were cut and diesel hauled mine-cars were employed, but the problems persisted to plague the pit. At Vesting Day five units were working, all on the west side of the pit, and all suffering from spontaneous combustion and difficult conditions.

The Beamshaw Seam, which had been unsuccessfully worked earlier in the life of the colliery, was opened up to the south. The seam was worked for a number of years, but again, difficult working conditions forced its closure in 1960. An attempt to work the Winter Seam was also unsuccessful.

In 1950 it was decided to sink boreholes in an attempt to test the viability of extending the coal faces in a safe and economic manner. Over forty boreholes were sunk. The borings proved that the colliery take was beset with faults and the thirty-four years of struggle to work the colliery were brought to an end. When the colliery was sunk, boring to test for strata conditions was not carried out to the same extent as in later decades. If boring had been carried out more vigorously it is unlikely that the colliery would have been sunk.

The death-knell for the colliery came on 20 May 1964 when a mysterious explosion took place in the Barnsley Seam. The seam was sealed in the interest of safety. Little coal remained elsewhere in the colliery and, after close consultation with the unions, the NCB took the decision to close the colliery. On 28 August 1964 the *Colliery Guardian* reported, 'Britain's Unluckiest Pit is Doomed: Upton Colliery to Close In a Few Months' In November 1964 the colliery finally closed.

Seams Worked:

Winter Seam
Beamshaw Seam
Barnsley Seam, 711yds deep, 14ft 9in thick.

12
Developments After Nationalization

Prior to nationalization there were a number of collieries, along the outcrop of the coal measures in the west, that worked the coal from drifts. These collieries included; Grange, Wentworth Silkstone, Stocksbridge, Smithywood and the early adit workings at Woolley Colliery. In addition there were a number of collieries working shallow seams, particularly the Shafton, which had started as shallow shaft mines and which were converted to drift mines around the time of nationalization. These include Goldthorpe and Highgate Collieries. Dearne Valley Colliery, which also worked the Shafton Seam, was sunk as a drift mine from its conception.

In the first half of the twentieth century the various colliery companies had made extensive use of underground cross-measure drifts to link the seams being worked and concentrate coal winding at a small number of seam horizons. This was particularly evident at collieries working a relatively large number of seams (for example four to six), or where seams were in close proximity to each other, making good sense to wind from a common level.

The principal advantage of a drift was the continuous flow of production that could be achieved, unlike a shaft which produced an intermittent flow (frequently the limiting factor to output at a colliery). In all cases it was possible to achieve a much higher output from a drift than a shaft. A drift also acted as a convenient 'buffer', and could hold many thousands of tons of coal 'in transit' to the surface.

Following nationalization the NCB set about driving drifts at a number of collieries. In 1947 the NCB started driving a deep drift at Wharncliffe Woodmoor Nos 4 and 5 Colliery. The colliery was uneconomic at vesting day and, to improve its viability, a steep drift was driven from the surface to connect with the deepest workings of the colliery. With a vertical depth of 410yds, the drift was probably one of the deepest ever driven by the NCB.

At Elsecar Colliery, between 1942 and 1943, a 1 in 5 gradient haulage drift was driven to the Haigh Moor Seam from the surface. This high productivity drift was further enhanced by the installation of a cable belt down the drift and along the main South Level, in 1956. The drift proved to be an efficient transport system to the surface and helped maintained the colliery's profitability.

In 1957 surface drifts were sunk to work the upper seams at North Gawber Colliery. In 1965-66 an underground drift, connecting into the intake drift of the original pair of drifts, was driven to access lower Seams. The new underground drift connected into the No.1 intake drift of the original pair of drifts. This enabled the whole of the output from the lower seams to be transported to the surface and dispensed with the coal-drawing shafts, which were filled soon after.

A major reconstruction scheme at Dodworth Colliery in the mid-1950s involved the elimination of long, underground rope-haulages and the installation of a cable-belt to convey coal to the surface via an existing drift. Later, in 1962, a surface drift was driven to the lower seams and coal from the Whinmoor and Silkstone Seams was conveyed to the surface by the new drift.

At Kiveton Park Colliery, a reconstruction scheme, completed in 1977 at a cost of over £2million, changed the colliery from a shaft mine to the first full-scale drift mine in South Yorkshire. The Jubilee Drift, named to commemorate the Queen's Silver Jubilee in 1977, linked to the Clowne Seam, some 200yds below the surface.

At Treeton Colliery a £20million scheme was approved in 1976 to transform the colliery and increase capacity. Between 1976 and 1981 a drift was sunk to reach the deeper reserves in the Swallow Wood Seam. Coal was transported up the drift to a staple shaft and along a shallow 900yd long tunnel, beneath the village of Treeton, to the coking ovens at Orgreave. By 1982 coal was flowing from the coalface to the coking plant at Orgreave, without seeing the light of day.

In the 1970s the NCB were sinking high-productivity drift mines, often in the pit yard of an existing colliery, or on the sites of closed collieries, to work the shallow upper seams using high-output retreat mining techniques. The lead-time to bring a drift mine into production was approximately two years, against up to ten years for a deep shaft mine. The reduced investment and speed of start-up of a drift mine weighed heavily in its favour, particularly in the volatile market place of the 1970s.

The pioneering site for this new 'breed' of super drift mine was Riddings Drift, a drift mine sunk in the pit yard of South Kirkby Colliery, which started production in June 1969. Though the unit shared the same surface facilities as South Kirkby it was operated as a completely separate unit, working the Shafton Seam and employing full retreat mining technology. The project was designed to produce coal highly competitively. The experience gained would prove to be valuable to the mining industry.

Following closely behind, and based on the highly successful system pioneered at Riddings Drift, was Royston Drift. Situated adjacent to the closed New Monckton Colliery, using some of the facilities of the former colliery, the drift was designed to work the three Sharlston Seams, which lay at shallow depth. Production at Royston Drift commenced in September 1976. The efficiency of Royston Drift was brought about by the application of the latest mining techniques, and retreat mining enabled them to set and break European and world production records. The mine was very successful, producing on average 1m tons per year. It became a victim of its own success when it faced exhaustion of its reserves and closed, in September 1989, after little more than a decade of production.

Between 1979 and 1984 the Barnsley Area of the NCB was subject to a massive reorganization based around three complexes, the West Side Complex centred on Woolley Colliery, the East Side Complex centred on South Kirkby Colliery and the South Side Complex centred on Grimethorpe Colliery. At each of these complexes coal from satellite collieries was transported underground to be brought up a drift into a central colliery for processing. At Grimethorpe the drift was planned for an eventual capacity of 3m metric tons per annum and coal was fed into one of the world's most advanced coal preparation plants.

In the late 1960s successful explorative borings near Barnburgh Main resulted in provisional plans for a drift mine, similar in concept to the pioneering Riddings Drift. The drift was to produce 10,000-15,000tons of good quality steam-coal per week. It was to be sunk from the pit yard at Barnburgh Main. At the same time a boring programme carried out in the Mexborough area also showed good potential, and a drift mine called Cresacre Drift was planned. Unfortunately neither drift was sunk. Collieries sunk in this period include Kinsley Drift, Ferrymoor/Riddings Drift and Royston Drift

Kinsley Drift Colliery

Located at Fitzwilliam, between the towns of Wakefield and Pontefract, the colliery was well-placed for access by both road and rail. The drift mine was developed on the site of the earlier Hemsworth Colliery, a shaft mine, which had closed in 1967, following its merger with South Kirkby Colliery. The site was on a section of Earl Fitzwilliam's estate. Only twelve years after mining ceased at Fitzwilliam it was possible to exploit the shallow seams beneath the area. Like Royston Drift, Kinsley was opened to exploit a parcel of unworked coal in a heavily mined area.

Preparatory work on the new mine began in April 1978. By the time coal production in the Shafton Seam commenced, in 1979, just fifteen months after cutting the first sod, the colliery had set a European record for the shortest time to get a deep mine into production. Two surface drifts, 1,300yds long, at a gradient of 1 in 4, were driven through the Shafton Seam to the Sharlston Yard Seam.

At a cost of nearly £20million, the new Kinsley Drift Mine was developed and officially opened on 18 August 1979, by Norman Siddall, the Deputy Chairman of the NCB. The new drift mine had a projected life of more than twenty years, at an output in excess of 0.5m tons per annum.

The bulk of the output was transported by rail direct to CEGB power stations, and the balance went to the multi-product washery at South Kirkby Colliery, where it was blended with coal from South Kirkby and Ferrymoor Riddings Collieries, before being sent on to CEGB power stations.

In 1985, after only six years, Kinsley was facing problems. In latter years the colliery suffered from a combination of unfavourable geological conditions and a poor quality product. Between April and the end of November 1985 the colliery lost £400,000, about £40 for every ton mined. At the end of 1985 the NCB put the colliery on the closure review procedure, with a recommendation that the colliery be closed. Kinsley Drift closed on 11 July 1986, with 283 men on the books.

Seams Worked:
Shafton Seam, 143yds deep, 5ft 11in thick.
Sharleston Yard Seam.

Ferrymoor /Riddings Drift

In June 1969 production started from Riddings Drift, a new drift whose entrance was sunk in the pit yard of South Kirkby Colliery. The new drift was 850yds long, and sloped at a gradient of 1 in 4.3, to the 200yd deep Shafton Seam. The drift was ventilated by the upcast at South Kirkby and shared the same surface facilities, but was operated as a completely separate unit. The drift worked the 54in. thick Shafton Seam, and was the first in Britain to work 'in the seam' retreat mining, using a shorter than average longwall face. The drift was linked to Ferrymoor Colliery workings, to enable Ferrymoor coal to be worked and transported more economically.

The drift had been designed to meet the objectives set by the coal industry in the mid-1970s, of producing coal competitively, in order to compete with rival forms of energy. The project cost £611,000, and was designed to work known reserves in the Shafton Seam, amounting to some 7m tons, employing retreat mining techniques.

The drift employed 160 men and in early 1970 it became the first British mine to produce coal at more than 8tons (162.5cwt) a shift for every man employed, this was almost one ton a man more than the previous record. The drift had a target of producing at a rate of ten tons for every man employed. The drift was worked entirely by 'in seam' mining methods using the latest high powered machines and short production faces, which were extensively studied in the United States.

On 1 April 1973 Riddings Drift and Ferrymoor Colliery, at Grimethorpe, merged to form Ferrymoor/Riddings Drift. The two collieries already shared common facilities and this was a natural progression.

In 1973 the Sharlston Seam was being developed, and the first face was expected to come into production in 1974.

In March 1987 Ferrymoor/Riddings merged with South Kirkby Colliery, and closed on 25 March 1988 when South Kirkby Colliery closed.

Seams Worked:
Shafton Seam, 4ft 6in thick, 181yds deep.
Sharlston Seam.

Royston Drift Mine

Royston Drift Mine was situated four miles north-east of Barnsley, adjacent to the closed New Monckton Colliery. It was born out of the NCB's 'Plan for Coal', approved by the Government in 1974. The plan promoted the need for expansion in the coal industry to meet the anticipated demand for coal in the 1980s, following the acceleration in closures during the oil glut of the 1960s. The strategy was to create new, more efficient mines and develop entirely new coalfields such as Selby. This strategy was to be brought about by massive investment over the next ten years, to yield over 40m tons of additional capacity. By the end of September 1976 the NCB had approved over eighty major projects designed to provide 26m tons of capacity by the mid-1980s, Royston Drift was one of these projects. The drift was to be based on the highly successful system pioneered at Ferrymoor Riddings, and the experience gained at nearby North Gawber Colliery.

Authority for the development of a 400,000tons per annum drift mine was given in 1974. The new mine had an anticipated life of ten years, and cost £4.5million. The mine was to be developed to maintain Barnsley Area output, by replacing output from collieries whose reserves were exhausted, and providing employment for men released by closing collieries. The development would improve Area productivity, particularly with respect to increasing output from retreat mining, and to work reserves which could not be worked by any other means. The target productivity rate for the new mine was seven tons per man-shift, more than three times the national average. The drift was expected to employ 230 men, producing coal exclusively for the power generation market.

Reserves of the colliery were contained in an approximately rectangular area of four miles by two miles, isolated by faults from existing collieries. A programme of boring in 1972-73 proved the Sharleston Top, Sharlston Low and Sharlston Yard Seams over an extensive area of

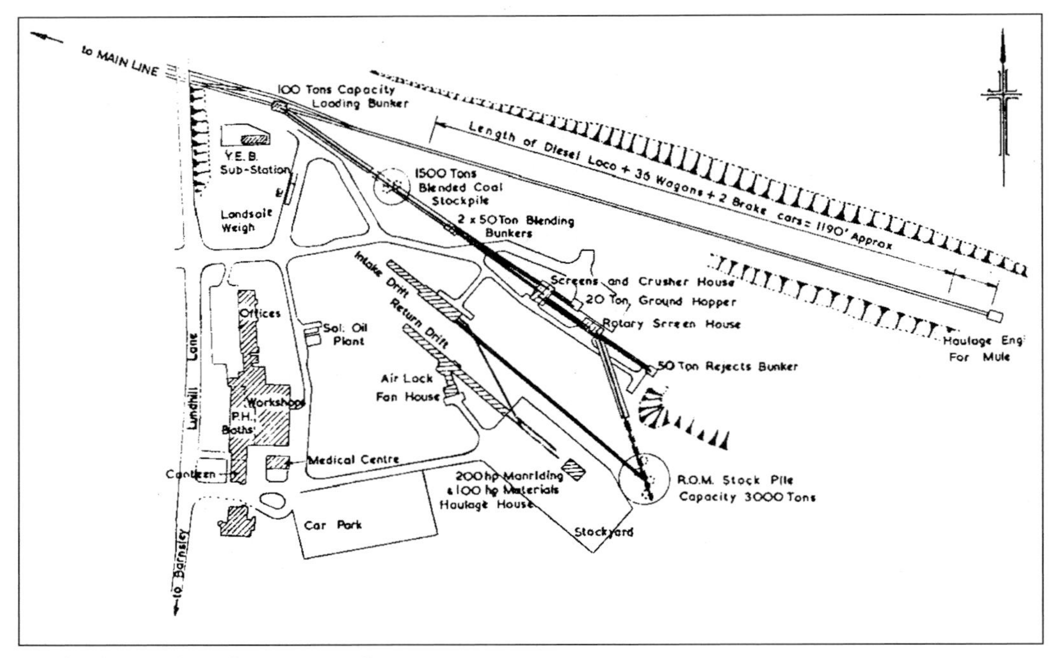

Surface layout plan of Royston Drift

the take. Reserves were calculated to be greater than 11m tons. The lower seams (250yds and deeper) in the area had been extensively worked and, although faulting was known to exist, it was not anticipated to be a major problem.

The three Sharlston Seams were at shallow depth, in the take of the abandoned Monckton No.6 Colliery. The Sharlston Top Seam lay at a depth of 200ft, in the east, and over 300ft, in the north and west. The Sharlston Low was 80ft below, and the Sharlston Yard was a further 70ft deeper. The three seams lay between depths of 70yds and 200yds.

Sinking commenced on 18 November 1974 with the construction of the portals, followed by simultaneous drivage of two 1 in 4 gradient cross-measure drifts in January 1975. As the drivage continued, increasing quantities of water were encountered. Seepage of between 100 and 300gallons per minute were encountered as the drifts were constructed. Despite the seepage the drifts were completed in twenty-five weeks at an average weekly advance of just over 16yds and a peak of 34yds. The two parallel drifts were 428 and 416yds in length to the Sharlston Seam.

The former New Monckton Colliery, which closed in 1966-1967, was located across the road from Royston Drift. In the late 1950s new baths and office buildings had been built and, although in a poor state of repair, were of sound construction. The bathing area was too large for the number of men employed at Royston Drift, so part of it was converted into a workshop and sub-station. Other buildings refurbished included the canteen, medical centre and office block. Access roads and the surface coal preparation plant were completed before the official colliery opening date.

Described by the *Yorkshire Post* as 'Britains first new pit for twelve years', the colliery was officially opened by the NCB Chairman, Sir Derek Ezra on 11 September 1976. The Sharlston

Low Seam was worked first. Face equipment installed during August 1976, commenced production on Monday 13 September 1976.

The reserves at Royston were to be worked in conjunction with the proposed Cawker Opencast Project. Output from the two sites was to be blended and loaded onto 'merry-go-round' trains for despatch to power stations. Opposition to the opencast site resulted in an abridged scheme, which produced 250,000tons.

The efficiency of Royston Drift was brought about by the application of the latest mining techniques. The extent of the coalfield was well understood, as was the position of the three major fault systems. The drifts were driven to the perimeter of the area being worked and the seam worked backwards towards the starting point, the opposite of 'traditional' mining. This approach became known as retreat-mining and, in conjunction with the use of double-ended drum-shearers, which cut coal in both directions, facilitated high-output, efficient mining. In addition, only coal was cut, with virtually no dirt, again enhancing efficiency. Retreat mining enabled Royston Drift to set and break European and world production records. These outputs were, however, achieved at a cost to the environment, as the Sharleston Seam was relatively shallow, which resulted in subsidence damage to roads, utilities and houses.

In the mid-1980s the colliery was working the Sharlston Low and Sharlston Yard Seams, and was the world record holder for coalface productivity. The system of working achieved consistent rates of high performance, including a record face output per man-shift of 362tons and a world record shearing run of almost two miles in a single shift.

The mine was very successful, producing an average of almost 1M ton per year, becoming a victim of its own success with the exhaustion of its reserves after only a decade of production. The colliery was closed on 8 September 1989.

Seams Worked:

Sharlston Low Seam, 93yds deep.

Sharlston Yard Seam

Bibliography

Main Sources

***Colliery Guardian*, Vols 62-233 (1891-1985)**

'An open and Shut Case', *Colliery Guardian*, Vol.211, 1965.

Brewer, N.C., 'Features of High Temperature Coal Carbonization in Yorkshire 1900-1962', *Colliery Guardian*, Vol.233, 1985.

'Britain's Unluckiest Pit is Doomed', *Colliery Guardian*, Vol.209, 1964.

'Britain's Unluckiest Pit is Doomed', *Colliery Guardian*, Vol.209, 1964.

'Coal winding stops at Wharncliffe Silkstone', *Colliery Guardian*, Vol.212, 1966.

'Coke Oven Plant at Hoyland Silkstone Collieries', *Colliery Guardian*, Vol.117, 1919.

'Colliery modernization at Newton Chambers Collieries', *Colliery Guardian*, 134.

'Difficult times in Britain's richest coalfield', *Colliery Guardian*, Vol.212, 1966.

'Hickleton Main in Trouble', *Colliery Guardian*, Vol.216, 1968.

'Kilnhurst Collieries', *Colliery Guardian*, Vol.149, 1934.

'Manvers Combined Mine', *Colliery Guardian*, Vol.194, 1957.

'NCB Reorganization', *Colliery Guardian*, Vol.214, 1967.

'North Gawber Washery', The *Colliery Guardian*, Vol.182, 1951.

Pitt, R., 'Which Plan for Coal?, An assessment of the period 1972-1984 and the future role for coal', *Colliery Guardian*, Vol.232, 1984.

'Proposals for a Combined Mine', *Colliery Guardian*, Vol.182, 1951.

'Riddings Practice Takes Record', *Colliery Guardian*, Vol.218, 1970.

'Robens visits Silverwood', *Colliery Guardian*, Vol.214.

Sinclair, J., 'Houghton Main Colliery – I', *Colliery Guardian*, Vol.146, 1933.

Sinclair, J., 'Manvers Main Collieries – II', *Colliery Guardian*, Vol.147, 1933.

'The Yorkshire Coalfield, Aldwarke Main & Carhouse Collieries, and Canklow Winning', *Colliery Guardian*, Vol.63, 1892.

'The Yorkshire Coalfield, Barrow Collieries', *Colliery Guardian*, Vol.63, 1892.

'The Yorkshire Coalfield, Cadeby Main Colliery', *Colliery Guardian*, Vol.69, 1895.

'The Yorkshire Coalfield, Cadeby Main Colliery', *Colliery Guardian*, Vol.76, 1898

'The Yorkshire Coalfield, Denaby Main Colliery', *Colliery Guardian*, Vol.63, 1892.

'The Yorkshire Coalfield, Fitzwilliam-Hemsworth Collieries', *Colliery Guardian*, Vol.78, 1899.

'The Yorkshire Coalfield, Grimethorpe Colliery', *Colliery Guardian*, Vol.78, 1899.

'The Yorkshire Coalfield, Hickleton Main Colliery', *Colliery Guardian*, Vol.76, 1898.

'The Yorkshire Coalfield, Hoyland Silkstone Colliery', *Colliery Guardian*, Vol.78, 1899.

'The Yorkshire Coalfield, Kiveton Park Collieries', *Colliery Guardian*, Vol.69, 1895.

'The Yorkshire Coalfield, Manvers Main Collieries', *Colliery Guardian*, Vol.26 February 1892.

'The Yorkshire Coalfield, Monk Bretton Colliery', *Colliery Guardian*, Vol.63, 1892.

'The Yorkshire Coalfield, Monckton Main Colliery', *Colliery Guardian*, Vol.62, 1891.

'The Yorkshire Coalfield, North Gawber Colliery', *Colliery Guardian*, Vol.74, 1897.

'The Yorkshire Coalfield, Nunnery Collieries', *Colliery Guardian*, Vol.68, 1894.
'The Yorkshire Coalfield, Rockingham Colliery', *Colliery Guardian*, Vol.69, 1895.
'The Yorkshire Coalfield, Rotherham Main Colliery (Canklow Winning)', *Colliery Guardian*, Vol.63, 1892.
'The Yorkshire Coalfield, Rotherham Main', *Colliery Guardian*, Vol.76, 1898.
'The Yorkshire Coalfield, The Nunnery Collieries', *Colliery Guardian*, Vol.76, 1898.
'The Yorkshire Coalfield, The Oaks Collieries', *Colliery Guardian*, Vol.63, 1892.
'The Yorkshire Coalfield, The Old Silkstone & Dodworth Coal & Iron Company's Collieries', *Colliery Guardian*, Vol.63, 1892.
'The Yorkshire Coalfield , Thorncliffe, Smithywood, Tankersley and Grange Collieries', *Colliery Guardian*, Vol.69, 1895.
'The Yorkshire Coalfield - Wath Main Colliery', *Colliery Guardian*, Vol.76, 1898.
'The Yorkshire Coalfield, Wharncliffe Silkstone Collieries', *Colliery Guardian*, Vol.63, 1892.
'The Yorkshire Coalfield, Wharncliffe Silkstone Collieries', *Colliery Guardian*, Vol.76, 1898.
'The Yorkshire Coalfield, Wombwell Main Colliery', *Colliery Guardian*, Vol.74, 1897.
'The Yorkshire Coalfield, Woolley Collieries', *Colliery Guardian*, Vol.74, 1897.
'Thin seam shearing at Wentworth Silkstone', *Colliery Guardian*, Vol.212, 1965.
'Upton Colliery 1, A new West Yorkshire undertaking', *Colliery Guardian*, Vol.136, 1928.
'Winding-engines at Manvers Main Colliery', *Colliery Guardian*, Vol.195, 1957.
'Yorkshire Division', *Colliery Guardian*, Vol.214, 1967.
'Yorkshire Main Colliery', *Colliery Guardian*, Vol.187, 1953.
'£1million Reconstruction at Kiveton Park', *Colliery Guardian*, Vol.209, 1964.

Colliery Engineering, Vols 6-35 (1929-1956)

'Brookhouse Colliery', *Colliery Engineer*, Vol.22, 1945.
'Carlton Main Colliery Group', *Colliery Engineer*, Vol.6, 1929.
'Developments in Yorkshire', *Colliery Engineer*, Vol.35, 1958.
'Drifting at Goldthorpe', *Colliery Engineer*, Vol.35, 1958.
'Economics in a Yorkshire Colliery Group', *Colliery Engineer*, Vol.6, 1929.
'Electrification of a Yorkshire Colliery', *Colliery Engineer*, Vol.11, 1934.
'Hatfield Main Colliery', *Colliery Engineer*, Vol.6, 1929.
'Houghton Main Colliery', *Colliery Engineer*, Vol.8, 1938.
'Kilnhurst Colliery', *Colliery Engineer*, Vol.11.
'Manvers Main Collieries Ltd – 1', *Colliery Engineer*, Vol.7, 1930.
'Manvers Main Collieries Ltd – 2', *Colliery Engineer*, Vol.7, 1930.
'New Stubbin Reconstruction', *Colliery Engineer*, Vol.36, 1959.
'New Stubbin Reconstruction', *Colliery Engineer*, Vol.37.
'South Kirkby Colliery, Yorkshire', *Colliery Engineer*, Vol.6, 1929.
'Thorne Colliery – II', *Colliery Engineer*, Vol.5, 1928.

Iron & Coal Trades Review, Vols 84-171 (1912-1955)

'Aerex fan at Dodworth', *Iron & Coal Trades Review*, Vol.133, 1936.
'Askern Colliery', *Iron & Coal Trades Review*, Vol.87, 1913.

Atkinson, J., 'Retreating Longwall Faces at Rockingham Colliery', *Iron & Coal Trades Review*, Vol.9, October 1953.
'Bullcroft Main Colliery', *Iron & Coal Trades Review*, Vol.125, 1932.
'Coal-face lighting from Mains at Nunnery Colliery', *Iron & Coal Trades Review*, Vol.127, 1933.
'Coal washing and preparation at North Gawber Colliery', *Iron & Coal Trades Review*, Vol.162, 1951.
'Developments at Cortonwood Colliery', *Iron & Coal Trades Review*, Vol.84, 1912.
'Electrification at Manvers Main', *Iron & Coal Trades Review*, Vol.160.
'Explosion at Barnsley Main Colliery', *Iron & Coal Trades Review*, Vol.145, 1942.
'Large endless rope haulage at Aldwarke Main Colliery', *Iron & Coal Trades Review*, Vol.142, 1941.
Machin, C., 'Houghton Main Colliery, Reorganization and Mechanization of the Parkgate Seam', *Iron & Coal Trades Review*, Vol.171, 1955.
'Manvers Main Collieries, The Barnborough Colliery', *Iron & Coal Trades Review*, Vol.125, 1932.
'New Simon Carves Coke Oven installation at Manvers Main Colliery', *Iron & Coal Trades Review*, Vol.128, 1934.
'Projection, Development and Working of Monk Bretton Colliery', The *Iron & Coal Trades Review*, Vol.152, 1946.
'Reconstruction of Barrow Collieries, Barnsley', *Iron & Coal Trades Review*, Vol.108, 1924.
'Reconstruction of Kilnhurst Colliery', *Iron & Coal Trades Review*, Vol.129, 1934.
'Some aspects of rapid development through stone and coal', *Iron & Coal Trades Review*, Vol.154, 1947.
'The future of Kilnhurst', *Iron & Coal Trades Review*, Vol.150, 1945.
'Vertical winding-engine at Wath Main Colliery', *Iron & Coal Trades Review*, Vol.117, 1928.
'Waleswood to close – 'Uneconomical' says the NCB', *Iron & Coal Trades Review*, Vol.156.
'Working the Shafton Seam at Highgate Colliery: Siskol Electric Machine Cutters', *Iron & Coal Trades Review*, Vol.133, 1936.

Mining Journal, Vols 42-50 (1872-1880)

'Collieries in Yorkshire, The Hoyland Silkstone, The finest shaft in England', *Mining Journal*, Vol.47, 1877.
'Collieries in Yorkshire, Cortonwood', *Mining Journal*, Vol.47, 1877.
'Ironworks and Collieries in Yorkshire: Manvers Main Colliery' *Mining Journal*, Vol.42, 1872.
'Ironworks and Collieries in Yorkshire: The Monk Bretton Colliery', *Mining Journal*, Vol.42, 1872.
'Opening out of a new coalfield by Lord Wharncliffe', *Mining Journal*, Vol.43, 1873.
'South Yorkshire Collieries: Messrs Newton, Chambers, & Co.'s Thorncliffe Collieries', *Mining Journal* (Supplement), 3 June 1871.
'The Coal & Iron Industries of Yorkshire: Monk Bretton Colliery', *Mining Journal*,Vol. 50, 1880.
'The Iron & Coal trades of South Yorkshire', *Mining Journal*, Vol.39, 1869.
'The Silkstone Coal Company's Collieries, Dodworth', *The Mining Journal*, Vol.43, 1873.

Supplementary Sources

Allott, F., notes in the possession of the National Mining Museum, 1935.
Barnet, A.L, *Railways of the South Yorkshire Coalfield from 1880*, (The Railway Correspondence & Travel Society, 1984)

Barrow Colliery Centenary 1876-1976, Open Day Handbook, 18 September 1976.

Bennett, A., *Rockingham Colliery Through the Ages* (private publication, 1979)

Booth, A.J., *A Railway History of Denaby & Cadeby Collieries*

Booth, A.J., *Industrial Railways of Manvers Main & Barnburgh Main*, Industrial Railway Society, 1996.

Brodsworth Main Colliery 1905-1955, Jubilee Celebration souvenir brochure.

Clamp, R., *New Monckton Collieries.*

Clarke, P. & Bradley, L., 'Rehabilitation of No.4 shaft – Barnsley Main', paper presented to the Barnsley & District Mining Society, 6 March 1984.

Coal News

Colliery Guardian Guide to the Coalfields.

Colliery Year Book & Coal Trades Directory.

Davies, P., *Dearne Valley Collieries.*

Department of Scientific & Industrial Research. *Fuel Research: Physical Survey of the National Coal Resources No.13, The Yorkshire, Nottinghamshire & Derbyshire Coalfield. South Yorkshire Area. The Parkgate Seam*, October 1928

Department of Scientific and Industrial Research. *Fuel Research: Physical Survey of the National Coal Resources No.18, The Yorkshire, Nottinghamshire & Derbyshire Coalfield. South Yorkshire Area. The Barnsley Seam*, March 1931.

Doncaster Evening Post, 28 September 1970.

Fox, W., *Coal Combines in Yorkshire*, (Labour Research Department, 1935)

Gibson, W., *Coal in Great Britain*, (Edward Arnold, 1920)

Guide to Sheffield, 1899.

Hall, A., *King Coal – Miners, Coal and Britain's Industrial Future*, (Pelican, 1981)

A History of the County of Yorkshire, The Victoria County Histories of England.

History of the Monckton Collieries and Coking Plant, National Mining Museum, Caphouse Colliery, Yorkshire.

Hopkinson, K., 'North Gawber Colliery, Not just a hole in the ground', notes.

Houghton Main Colliery, RCHME report, NBR No.91140, undated.

Houghton Main Colliery Jubilee Open Day, 16 July 1977.

Houghton Main Colliery, Outline of case for conserving this colliery complex, initialled M. O'B. / D.T.D., 18 June 1993.

Hunter, *Draft report on the future of Markham Main*, November 1939, (copy in Sheffield Archives).

Jones, J. & M., '...A most enterprising thing...' An illustrated history to commemorate the 200th anniversary of the establishment of Newton Chambers at Thorncliffe.

Journal of the National Mining Memorabilia Association, Vol. 2

Kiveton Park Colliery Family Day Booklet, July 2 1978.

Littleby, R., *A Short Illustrative History of the Rothervale Collieries.*

Malpass, J., *The Building & Development of Hickleton Main Colliery.*

Maltby Colliery, British Coal, South Yorkshire Area, October 1988.

Maltby Colliery Reconstruction, Qualter, Hall & Co., Barnsley, 1960.

Maltby Main Colliery, Golden Jubilee Brochure 1911-1961.

New Stubbin Colliery Family Day, NCB Public Relations (Yorkshire), 17 July 1977.

Price, G., *The Midland (Amalgamated) District (Coal Mines) Scheme 1930*, Report to the Executive Board, 6 February 1942.

Rawnsley, M., Notes on North Gawber Colliery, 1970.

Report on the Explosion at Houghton Main Colliery, Yorkshire, June 1975, Health & Safety Executive.

Rossington Colliery, RCHME.
Rockingham Colliery Centenary Open Day Booklet, 31 March 1975.
Rossington, T., *The Story of Treeton Colliery 1875-1975*, Metropolitan Borough of Rotherham Libraries, 1976.
Sections of Strata of the Yorkshire Coalfield, Midland Institute of Mining Engineers, 1927.
Sheffield Telegraph, 21 April 1962.
Silver Jubilee of Their Majesties King George V and Queen Mary, 6 May 1935.
Smithson, H.F., *Report on Mitchell Main & Darfield Main Collieries*, 19 February 1931.
Stables, R.C., Area Surveyor & Minerals Manager, NCB Barnsley Area, letter, 12 July 1977.
Statement showing outline of History and Development in the South Yorkshire Coalfield, The South Yorkshire Coal Trade Association, 1925.
Stephenson-Clarke, C., *Notes on the Coals and Coal-seams in The Midland Coalfield (West Yorkshire, South Yorkshire, Nottinghamshire and Derbyshire)*, August 1944.
Success in South Yorkshire, NCB.
Success in South Yorkshire: Maltby Colliery, NCB.
Swift, W., *History of Darfield Main Colliery 1860-1900*, Barnsley, 1990.
The Bullcroft Main Colliery – General description.
The Engineering Resources of Markham & Company (booklet).
The John Goodchild Collection, Wakefield, Yorkshire.
The North Country and Yorkshire Coal Annual, 1924.
The Nunnery Colliery Company Limited, Sheffield and Rotherham (Illustrated), 2000.
Thorpe, P.A., *Monckton its origins and history*, 1997.
Transactions, National Association of Colliery Managers, 6.
Treeton Times, final edition.
Visit to Houghton Main Colliery, Thursday 8 July 1954, Institution of Mining Engineers Summer Meeting, 1954.
Visit to North Gawber Colliery, Friday 9 July 1954, Institution of Mining Engineers Summer Meeting, 1954.
Wath Colliery Family Day, NCB Public relations, 4 July 1976.
Whitelock, G.C.H., *250 Years in Coal, The History of Barber, Walker & Co. Ltd, Colliery Proprietors in Nottinghamshire & Yorkshire.*
Wilkinson, A., *A History of South Kirkby*, South Kirkby & Moorthorpe Town Council.
Woolley Colliery 1869-1969, NCB Barnsley Area.
Yorkshire Evening News.
Yorkshire Evening Post
Yorkshire Main Colliery, 19 November 1953.

Supplementary Articles

Albert, J, 'Deepening a shaft at Dodworth Colliery', *Journal of the Barnsley & District Mining Society*, Vol. 7
'Aldwarke Main Colliery', *Transactions*, Federated Institute of Mining Engineers, Vol. 3
'Barrow Colliery', *Transactions of the British Society of Mining Students*, Vol. 15
'Bentley Colliery', *Transactions of the Institution of Mining Engineers*, Vol. 50, 1915-1916
'Brodsworth Main & Bullcroft Main Collieries,' *Transactions of the National Association of Colliery Managers*, Vol. 14, 1917
'Bullcroft Main Colliery', *Transactions of the National Association of Colliery Managers*, Vol. 14, 1917.

'Cadeby Main Colliery', *Transactions of the Federated Institution of Mining Engineers*, Vol. 18, 1900.
'Cadeby Main Colliery', *Transactions of the South Wales Institute of Engineers*, Vol. 27, 1910-1911.
'Concentration of coal manipulation at Grimethorpe Colliery', *Transactions of the National Association of Colliery Managers*, Vol. 40, 1943.
Crane, D., 'Shaft work for the development of Parkgate and Thorncliffe Seams at Bentley Colliery', *CLS/CMS*, Spring 1983
'Discussion on improvements in winding appliances', *Transactions of the South Wales Institute of Engineers*, Vol. 21, 1898-1899.
'Electrification of Nunnery Colliery', *Transactions of the National Association of Colliery Managers*, Vol. 21, 1924.
Gray, G.D.B., 'The South-Yorkshire Coalfield', *Geography*, Vol. 32, 1947.
'Linking & restructuring of the electrical plant at a group of large collieries & works', *Transactions, of the National Association of Colliery Managers*, Vol. 34, 1937.
'Living on Coal', *South Yorkshire Times*, 8 August 1959.
Macfarlane, J., 'Denaby Main: a South Yorkshire mining village', *Studies in the Yorkshire Coal Industry*, Benson, J. & Neville, R.G., eds.
Oates, J.A., 'Coal Preparation Plant at Dodworth', *Transactions of the Barnsley Mining Students Society*, Vol. 12, 1961-1962.
'Pit in production 50 years and still smashing records', *Rotherham Advertiser*, 22 July 1961.
'Pit ponies versus compressed air haulage', *Transactions of the National Association of Colliery Managers*, Vol. 7.
Pyne, R., 'Royston Drift Mine, NCB Barnsley Area', *Colliery Guardian Annual Review*, August 1977.
Schumaker, E.F., 'The Divisions in 1956, A Historical Review of the Midland Institute of Mining Engineers', *Schumacher on Energy*, Kirk, G., Ed., (Abacus, 1983)
'Skip-winding at Barnborough Main Colliery', *Transactions of the National Association of Colliery Managers*, Vol. 41, 1944.
'Solid Stowing on a Mechanized Face', *Transactions of the Barnsley Mining Student's Society*, Vol. 12, 1961-1962.
'Steep workings at Kilnhurst Colliery', *Transactions of the National Association of Colliery Managers*, Vol. 46, 1949.
'Story of Reorganization of Magnitude and Complexity', *South Yorkshire Times*, 6 June 1959.
'The Frickley Colliery of the Carlton Main Colliery Co. Ltd', *Transactions of the National Association of Colliery Managers*, Vol. 15, 1918.
'The Pits of Yorkshire: Dodworth shows the way with machines', *Sheffield Telegraph*, 8 March 1962.
'The pumping appliances used in the sinking operations at the Cadeby new winning', *Transactions of the Federated Institution of Mining Engineers*, Vol. 3, 1891-1892.
'The testing of a Steart fan at Grange Colliery, South Yorkshire', *Transactions of the Institute of Mining Engineers*, Vol. 76.
Thompson, A., 'Maltby Main Colliery', *Transactions of the Institution of Mining Engineers*, Vol. 44
'Thrybergh Hall Colliery', *Transactions of the Federated Institute of Mining Engineers*, Vol 3
'*Visit of Headmasters to Brookhouse Colliery*', 25 April 1957.
'Visit of Inspection to the Warren House and Rotherham Main Collieries', *Transactions of the National Association of Colliery Managers*, Vol. 12.
'Visit to the Silverwood Pit of the Dalton Main Collieries', *Transactions of the National Association of Colliery Managers*, Vol. 9, 1912.
'Yorkshire Pit (South Elmsall Colliery)', *Coal*, January 1958.

Further Sources

Handbook for the Visit by HM the Queen and HRH the Duke of Edinburgh to Silverwood Colliery, Rotherham, Wednesday 30 July 1975.
Notes and records in the possession of the National Mining Museum, Caphouse Colliery, Yorkshire.
Records of Hathorn Davey & Co., Leeds City Libraries, Archives Department.

Reports and notes on visits by the Mines Department, Technical Adviser – South Yorkshire District:
Askern Main, visit particulars dated 27 January 1942.
Barnsley Main, additional particulars dated 2 April 1942.
Bentley, visit particulars dated 27 January 1942.
Darfield Main, visit particulars dated 1942.
Darton, visit particulars dated 19 January 1942.
Dearne Valley, visit particulars dated 18 March 1942.
Elsecar Main, visit particulars dated 20 March 1942.
Haigh, visit particulars dated 19 January 1942.
Kiveton Park, visit particulars dated 26 February 1942.
North Gawber, visit particulars dated 19 January 1942.
Orgreave, visit particulars dated 6 March 1942.
Treeton, visit particulars dated 6 March 1942.
Waleswood, visit particulars dated 24 February 1942.
Wharncliffe Woodmoor 1,2 & 3, visit particulars dated 20 January 1942.
Wharncliffe Woodmoor 4 and 5, visit particulars dated 20 January 1942.
Wombwell Main, visit particulars dated 24 February 1942.
Woolley, visit particulars dated 12 January 1942.
Report on Aldwarke Main, February 1942.
Report on Barnborough Main, 11 March 1942.
Report on Brierley, 3 March 1942.
Report on Brodsworth Main, 30 January 1942
Report on Bullcroft Main, 1942.
Report on Cadeby Main, 14 January 1942.
Report on Cortonwood, 16 February 1942.
Report on Denaby Main, 15 January 1942.
Report on Dinnington Main, 7 January 1942.
Report on Frickley, 31 March 1942.
Report on Grimethorpe Main Colliery, 16 February 1942.
Report on Hatfield Main, 9 March 1942.
Report on Hickleton Main, 23 January 1942.
Report on Maltby Main, 9 January 1942.
Report on Manvers Main No.1 and No.3 pits, 16 March 1942.
Report on Markham Main, 6 January 1942.
Report on Mitchell Main, 12 February 1942.
Report on Rossington Main, 10 January 1942.

Report on Rotherham Main, 2 March 1942.
Report on New Stubbin, 20 March 1942.
Report on visit to Thurcroft Main, 18 March 1942.
Report on Yorkshire Main, 1942.

NCB Archives, Sheffield Archives, Sheffield, South Yorkshire.
NCB records held by the Coal Authority at Berry Hill, Mansfield
NCB records held at the National Mining Museum, Caphouse, Yorkshire.
NCB Doncaster Library & Information Service Notes.

Various NCB records, notes and other archive material have been useful, including:
Barnsley Area, notes, 1979.
Booklet for visit of Imperial Defence College to Yorkshire Main Colliery, Doncaster, Monday April 16 1956.
Booklet from visit to S44s retreat Unit, Friday 28th April 1978.
Brief for visit of Sir Derek Ezra, NCB Chairman, 23 August 1979.
Brodsworth Colliery, notes, August 1978 and August 1979.
Darfield Main Colliery, notes
Dinnington Main Colliery, notes, 8 July 1964.
Doncaster Area, notes, February 1963, July 1963, February 1973, July 1973 and August 1983.
Elsecar Main colliery, notes, 1960s.
Frickley/South Elmsall Colliery, notes, August 1983.
Hatfield Colliery, notes, August 1982.
Houghton Main Coking Plant, letter, dated 27 March 1950.
Kinsley Drift Mine (handbook).
Maltby Colliery, notes, 8 July 1966.
Markham Colliery, notes dated August 1983 and March 1987.
North-Eastern Division, press releases, 22 April 1948 and 24 November 1949.
Press release, 20 April 1948.
Rockingham seam plans and reports.
Rossington Colliery, notes, August 1983.
South Elmsall Colliery, notes, May 1961.
South Yorkshire Area, data card.
Thorne Colliery Project Visit, report, 11 May 1981.
Yorkshire Regional Public Relations Department, May 1967.

British Coal, South Yorkshire Area, notes, April 1989.
British Coal, South Yorkshire Area, notes on Dinnington Colliery, March 1987.
British Coal, South Yorkshire Area, notes on Maltby Colliery, March 1987.
British Coal: South Yorkshire Area, notes on Rossington Colliery, March 1987.

Index

About the Author

Alan Hill was born in Cumbria in 1947, the year when the British coal industry was nationalised. At the age of seven Alan moved to a mining village in South Yorkshire where his father, and later his brothers, worked in the coal industry. At the age of nineteen he left South Yorkshire to pursue a career in engineering and moved to Birmingham, which has been his home for over thirty years, where he lives with his wife Carole.

Brought up in an environment where 'coal was king' and the remains of past industries abounded, it was not surprising that he developed a strong interest in industrial archaeology and in particular the history and technology of the coal industry. A keen photographer, over the years he has photographed collieries and their remains, compiling a large photographic collection.

This book is the result of a long interest in the South Yorkshire Coalfield and may hours of research, and is the author's fourth book related to the coal industry.